About the Author

Known to Internet weather enthusiasts around the globe as the Weather Doctor, Keith C. Heidorn, PhD, has studied weather and climate throughout his life. Dr. Heidorn is a graduate of the University of Michigan and also studied at the University of Guelph. After a career spent working for the Ontario provincial government and in private consulting, he now focuses his energies in public education on myriad weather and climate topics, ranging from the origin of snowmen to the science of thunder. In addition to maintaining his Weather Doctor Web site, he was a regular contributor to *The Weather Notebook*, a radio program aired throughout the United States, has taught applied climatology at the University of Victoria, and was a charter member of the Canadian Meteorological and Oceanographic Society's School and Popular Education Committee. A Great Laker by birth, Dr. Heidorn now resides in Victoria, British Columbia.

To Kristi and Derek, who put up with a weird dad who made them watch thunderstorms for fun.

AND NOW . . .
THE WEATHER

With The Weather Doctor
Keith C. Heidorn

FIFTH
HOUSE

Cover design and illustrations, and interior design by Cheryl Peddie/Emerge Creative
Cover photographs by Jose Luis Pelaez/CORBIS (Snowstorm in the City) and
William James Warren/CORBIS (Lightning Storm)
Edited by Alex Frazer-Harrison
Copyedited by Lori Burwash
Proofread by Lesley Reynolds

The publisher gratefully acknowledges the support of The Canada Council for the Arts and the Department of Canadian Heritage.

Canada Council   Conseil des Arts
for the Arts   du Canada

We acknowledge the financial support of the Government of Canada through the Book Publishing Industry Development Program (BPIDP) for our publishing activities.

Printed in Canada by Friesens
05 06 07 08 09/ 5 4 3 2 1

First published in the United States in 2006 by
Fitzhenry & Whiteside
121 Harvard Avenue, Suite 2
Allston, MA 02134

National Library of Canada Cataloguing in Publication Data
Heidorn, K. C.
And now... the weather / with Keith C. Heidorn.
Includes bibliographical references and index.
ISBN 1-894856-65-1
1. North America--Climate. 2. Weather. I. Title.
QC982.8.H45 2005   551.697   C2005-902362-7

Fifth House Ltd.
A Fitzhenry & Whiteside Company
1511, 1800–4 Street SW
Calgary, AB T2S 2S5
1-800-387-9776
www.fitzhenry.ca

Contents

Introduction

Weather has always been part of my life. I know it is a part of everyone's life—we have yet to have any human be born and spend their entire life on the moon—but, for me, it has played a very special part. Some of my few remaining memories of preschool involve weather. In grade 4, I did my first science fair project on weather, creating weather instruments out of old milk cartons, eyedroppers, clothes hangers, and cardboard—back in the days when science fair projects could be made for nickels. I didn't win, but I did get an honorable mention in the district.

My love of storm watching as a youth and into my teen years spawned a curiosity in meteorology that pointed me toward a career in the sciences, after briefly flirting with the idea of being a sports writer. After a few decades working as an applied atmospheric scientist, I decided to combine my early love of weather with my second love: writing. Today, I find myself constrained only by the local climate and the roll of the seasons.

These days, weather has become even more a part of my life as I have the time to watch weather systems slowly unfold and to delve into the types of weather stories that don't make the media reports or history books. The media, of course, jumps for the big sensational story—for example, competing for who can headline the most rainfall or the lowest windchill conditions—usually neglecting the weather whys. However, this treatment often puts a negative spin on weather, leaving the good news to the local weathercasters, who continue to trumpet "another great, sunny summer day ahead," even in the midst of drought and the threatening consequences thereof.

In 1997, when El Niño was making international headlines and blamed for everything from presidential headaches to a minister's foul-ups, I decided it was time for a different approach. I began designing a weather Web site I initially called Weather Eyes, until a colleague from the British Columbia government introduced me at an environmental open house as the Weather Doctor. And so my site had a new name. I intended a Groundhog Day 1998 rollout of the Web site because I have always had an affinity for that date. The great ice storms in Quebec and New England that January, however, pushed the premiere forward. In a case of serendipity, the main article of the new Web site was to be on ice storms, and it was written, ready to go when that big one hit. It was not a case of accurate long-range prediction, just a bit of serendipity.

More than seven years later, the site is growing steadily. I have used it to discuss topics from the influence author/artist Eric Sloane had on my interest in weather, to the cause of mirages, to a discussion of snowmen. My Weather Almanac holds nearly one hundred essays with a good mix of scientific, historic, and "artistic" viewpoints.

Most weather books focus on severe weather and leave out the hows, whys, and joys of everyday weather occurrences. Others with a more general bent follow a textbook-like treatment beginning with general principles and then covering related topics in individual sections. In my storyline, I follow the calendar and seasonal cycles and cover topics that are very familiar, as well as those that are often overlooked as we rush through our daily lives. I do admit to a 40s-latitude bias as I have lived my entire life within that northern hemisphere 40° to 50° latitude belt. This region is among the most interesting, having the full expression of seasonal differences that other global belts often miss. It is the prime home of the *Prevailing Westerlies*, the polar front, and the polar jet stream, so weather changes are often rapid and extreme. The old saying "If you don't like the weather, wait ten minutes" finds expression just about everywhere in this belt, though most locals believe it is unique to their area alone.

"Change" is the prime weather word here. A very close second must be "anticipation." I once heard Canadian author W. P. Kinsella, who wrote *Shoeless Joe* (made into the movie *A Field of Dreams*), describe the allure of baseball as being that it was a game of anticipation, whereas he deemed ice hockey, basketball, and football games that derived their allure through action. As a lifelong baseball fan, I agree with Kinsella and apply the same analogy to the weather.

Consider a ball game where the inning places Barry Bonds in the on-deck circle. As a baseball fan, you play out a variety of scenarios, all in anticipation of Bonds's at-bat with the hope, unless you're a fan of the opposing team, that he might hit a prodigious home run, one you can talk about for years, telling your grandkids that you saw Bonds hit that mighty home run.

Now apply that analogy to everyday weather. The five-day forecast suggests a blizzard or severe snowstorm may hit your region at the end of the forecast period. If you are a true weather fan, those same anticipatory juices start flowing, and you hope it will be a big one that you can talk about for years, telling your grandkids about the big storm of "Aught Five."

I am a confirmed weather watcher, a weather aficionado (but *never* a "weather weenie") because I love the anticipation, I love the chase, and I love the action when the winds blow and the precipitation falls, or when the sun shines and the cumuli grow into towering cloud men. I also love to know why what I sense and experience occurs.

That is where this book comes in. I will take you on my path of curiosity about many aspects of the weather while seeking the science behind them. Leaving my usual meat-and-potatoes balance between poetry, mythology, personal experiences, and science, I will focus mostly on the science. Not that this is a dry, textbook treatment—I use the poetry and feelings as spice (steak sauce), rather than as balancing companion (baked potato).

Many naturalists start their books in spring, when new life emerges after a winter's sleep. I think it more appropriate to follow the weather by following the sun, since it is the driving force behind all our weather conditions. Therefore, I begin with the *winter solstice*, the low point in the solar year. This date is the middle of solar winter and mostly an astronomical event. The media has a fixation on calling it the "official" start of winter, but that is highly misleading and an insult to anyone living north of 45° north latitude.

Before I get into the science themes, I hope to instill in you some of my joys in weather watching, including its stimulation of the senses as well as the intellect. My goal is to help you train your "weather senses" so that you become more aware of the conditions surrounding you and so you can begin to enjoy the beauty and joy of weather, rather than dread its expression. Our lives are too filled with fear and dread as it is.

Come, join my journey through the year—and always keep those weather eyes up!

The Joys of Weather Watching

When I sat down this morning to write, the sun was just breaking the horizon. The most brilliant crimson sky I had seen in many months—accentuated by variously shaped clouds—announced dawn's arrival. I stepped out onto the balcony to get a better view and inhaled the crisp morning air, clean with just the hint of sea. Needless to say, the event delayed my work briefly—too short a time, for the breathtaking dawn lasted but a quarter hour. Had it lasted two hours, I still would have stopped to watch. I am a compulsive weather watcher. It is in my genes.

I have been a weather watcher most of my life—more than fifty years. My first weather memories are from preschool days—basic childhood reactions to thunder and lightning. I have made my love for weather both a recreation and a vocation, describing weather through poetry as often as through equations and statistics. What I would like to convey to you in this chapter is not the science or the weather whys. We'll deal with that later. Instead, I want to help you develop your weather eyes.

Weather and various atmospheric phenomena are the most sensual aspects of life. We see weather, we hear weather, we smell weather, and we occasionally taste weather. We definitely *feel* weather. We often stop listing the senses at five, but other physical senses also respond to the weather, as do many mental and spiritual senses. I combine them all under the umbrella of having a "weather eye," being sensually aware of the weather around us.

Mark Twain is quoted as saying "Everybody talks about the weather," although it appears his collaborator on *The Gilded Age*, Charles Dudley Warner, actually wrote the statement. I have no doubt that weather is on everyone's mind, but weather is more than a subject for casual conversation. We live in weather daily. It surrounds us, it influences us, it controls our past, present, and future in more ways than we care to admit. Weather shapes our culture, our national and regional characters, our language, music, art, and literature, even our mental and physical health. It awes us with its power and then captivates us with its delicate beauty. And, sometimes, it rains on our parades.

The Pastime of Weather Watching

Weather watching rivals birding and astronomy as the top nature activity. What other activity can be done almost everywhere, at any time, by anyone? Unlike astronomy, weather watching can be done day and night. Bird watching can be done in a lot of places, but watching birds from high-rise apartments and offices or while flying in a commercial aircraft is not an easy task. But weather is right there for us to savor.

I have practiced weather watching mostly as a solo activity, but that does not preclude it from being a rewarding social affair—whether you're watching with another individual, the family, or part of a group. I have watched weather from a sickbed and from a mountaintop after an exhilarating climb. I have watched weather in the big city and in the northern Ontario wilderness. I have shared my weather stories with family and friends around the dining table and around the campfire.

Have I got you hooked yet? I hope so. Not only is weather watching fun and educational, it can also be the most inexpensive of pastimes and has the least impact on the environment of any activity I can think of. There are no age restrictions and no minimum level of physical ability.

I'll let you in on a little secret: you can do a whole bunch of other fun activities in combination with weather watching. For example, if you're interested in writing or visual arts, you can combine them with weather watching. Weather has been a great inspiration for many poets. Artists, whether their medium is pencil, pastels, watercolors, oils, or photography, can focus on the weather or use it to enhance the atmosphere surrounding a work's focal point.

Getting Started

I was going to begin by saying all you need to begin weather watching is two good eyes and a place open to the sky. After a moment's thought, I dropped the need for two good eyes. You can "watch" weather with less than two good eyes, and I can say with certainty that sight is not even necessary. Even I with two good, albeit myopic, eyes have often enjoyed weather watching using only the senses of hearing and feeling.

A "place open to the sky" is a location that allows the senses to link with the sky. By using senses other than vision, we can practice weather watching in a variety of unlikely locations that may lack visual contact with the sky. I have worked in offices where I was cut off from, or could not easily see, the sky, but there was usually some clue that announced to my weather eyes what was going on outside.

Stratus clouds and fog surround a mountain lake outside Eugene, Oregon.
(KEITH C. HEIDORN)

One of these was an air vent. It changed its sound under a variety of weather conditions. Wind and rain would send the weather-guard flap singing a song unique to a particular condition—whistling with different pitches as the wind speed changed, plinking and plunking as the rain changed intensity. There was even a sound peculiar to the striking of freezing precipitation (freezing rain, sleet, ice pellets): *clink.*

For our first foray into weather watching, let's focus on the visual aspects of weather. If possible, move to a position where you can see the sky. It doesn't matter if it is day or night.

Focus your attention on what you see out there. Are clouds visible? Is the wind blowing? Is it raining or snowing? If it is winter, is there snow on the ground? Pick one aspect of what you see and let it engulf your attention. What do you see that you may have missed before?

I heard someone back there say, "Keith, how can I see the wind? Air is invisible." That's a good question. A perceptive weather watcher sees not only the directly visible, but also the indirect indications of weather. Look again. Is there a tree or flag or laundry dancing in the wind, or litter or dead leaves skipping down the street? Since wind is air in motion, at times you can actually see the wind as it affects light passing through it. Look for a twinkling star or the shimmer of a distant streetlight. These are often manifestations of the wind that you can see.

Did you see something you hadn't noticed before or see it with a new appreciation? I hope so. Maybe you are lucky and see a storm roll in or out, a rainbow or halo in the sky, or lightning streak across the horizon. If not, look for something you may have heard about on the weather report or read about. While you may not find exactly what you are looking for, something fascinating will be visible. As you sharpen your weather eyes, you will even find interesting features in a seemingly uniform cloud layer or in dense fog. Look toward light sources, not just the sun or the moon, to provide interesting color variations. The times around dawn and dusk provide fascinating opportunities because bright reds, oranges, and yellows often dominate portions of the sky.

Once you have exercised your vision, try using your hearing, tactile feeling, or smell to scan the current weather condition. Gather in the *feel* of the senses. Now think about how the weather makes you feel emotionally and spiritually. Joyful? Depressed? Afraid? Apprehensive? Awed? Excited? If you continue such exercises for several minutes each day, I guarantee that after a while you will want to spend more time and delve deeper into weather watching.

Going on a Weather-Watching Field Trip

As you practice, you may have ventured from home and noticed the weather as you walked to work, went for a walk or run, or worked in the garden. This time, I recommend going farther afield, away from an urban area, for the sole purpose of watching weather. If that's not possible, seek out a destination where the sky view is minimally obstructed, such as a viewpoint from a tall building.

My first recommendation is to wear proper clothing. While shivering is one way to sense the weather, it isn't a lot of fun. I am a great advocate of dressing in layers. It works in all types of weather.

Now that you are dressed appropriately and out the door, where should you go? That depends on where you live. I live on southern Vancouver Island, so it is easy to go to the shoreline or a park at higher elevations or even a low-level park that gives me the opportunity to see around the compass, more than I would see from a backyard or apartment commons. No matter where you are, it is important you choose a place where you can see more sky than at home and be physically immersed in the weather. Let it surround you. Weather watching when walking or even running can be a fun combination of activities, but today we will stand or sit in one spot for a while.

Now repeat our earlier exercise. Look for weather phenomena you may or may not have experienced before. Soak in the pageant of the weather moving around you. Perhaps you did a little homework and went

The first formation of winter ice paints a delicate pattern. (KEITH C. HEIDORN)

7

out on a particular day because of atmospheric conditions that were unfolding—you're learning quickly. If not, don't worry about it; I'm a big fan of serendipity and enjoying what you find. Something about the weather will grab your attention, even if the sky is completely clear.

As you develop your weather eyes over the next year, look for a variety of areas where you may enjoy different weather phenomena. I go to the seashore because, in addition to the wave action, it offers unobstructed sky vistas where cloud and optical events are more visible. I can also enjoy the interaction between mountains and sky. In autumn, I look for a spot where trees are dropping leaves and the wind is freer to flow. Then I sit back and listen to the leaves chattering and watch them dance. I once worked near a park across from a glass-clad high-rise. I enjoyed spending lunchtime in that park, watching the sun and clouds reflecting off the glass. Ponds and lakes can do similar interesting things with light.

You may even want to go out in "bad" weather. If dressed appropriately, you can open your weather eyes to many sensual experiences—the taste of snowflakes, the beat of rain, the force of the wind pushing against your body, or the electricity in the air before a thunderstorm. However, please use common sense regarding your safety, especially during severe thunderstorms, lightning, or extreme cold. Storm chasing can be fun, but it can also be extremely dangerous.

Some Things to Look for in the Sky

I hear you asking "What should I be looking for?" I have purposely avoided any answer to this question, partly because it depends on where or when you are reading this, and the weather situation I may tell you to look for may not happen. In addition, I would like you to first get accustomed to the parade of regular weather conditions where you live. Learn the nuances of your neighborhood weather. Do clouds form over mountain peaks in the same way every day at the same time? Do sunset features have any correlation to the day's weather? Does the wind shift in a common way on particularly hot days? How do storms roll in and how do they play out?

Keep notes of your observations in the beginning and see if there are patterns. If so, maybe you can predict that some weather phenomenon will happen whenever some particular pattern occurs. Find that a particular pattern occurs under common conditions and you have undertaken some basic scientific research. No matter that others may have determined the same sequence years ago, it is still *your* discovery. Use this knowledge to look for the phenomenon in the future, and you will be a weather forecaster.

One allure of weather watching is the anticipation of what is to come. A forecast of a major storm can catch us in anticipation for several days. During a thunderstorm, watchers eagerly await the next flash of lightning and peal of thunder. If they match our expectations, we ooooh and aaaahhhh like an audience at a fireworks display. However, weather watching offers not only a high degree of anticipation, but also a wide variety of engrossing experiences once the event unfolds.

When you become an avid weather watcher, the weather becomes more than something to complain about. It becomes a treasure chest of potential memories: brilliant rainbows, rolling thunder, crisp mornings, fire in the sky, lacy snowflakes, and powerful winds. If you aren't presently a weather watcher, join me; start today. Develop those weather eyes! You won't regret it.

Sunset over the Pacific Ocean, at Wickaninnish Island, near Tofino, British Columbia. (KEITH C. HEIDORN)

Two

The Wheel of the Year

In Western civilization, the start of the calendar year is 1 January, New Year's Day. However, this is but a manner of convenience, for the year is a circle, an orbit, a wheel. It is, perhaps, a little more oval than circular; nevertheless, it is a path without beginning, without end, and having no true starting point. Wherever we start the year, we return to that place at the end of one cycle.

The ancients who venerated the sun placed their beginning at some point along its annual hill-and-valley trek through the sky that made sense to them. As they watched the sun rise and fall through the year, few understood that the variation came from the cockeyed tilt of our planet as it spun its daily-cycle circle round old Sol, but they found ways to plot the solar journey. They left us inscrutable megaliths such as Stonehenge, which are likely stone calendars set to remind their builders of the exact date of the solstices and/or equinoxes.

For millennia before hours were first struck and we began parsing time into minutes and seconds, some cultures subdivided the year into logical divisions: first halves, then quarters, and finally, like spokes of a great wheel, eighths. Others chose the moon's waxing and waning as a logical way to parse the year.

Defining the Spokes

The halves and quarters constitute what we today call *seasons*—initially, two seasons were likely recognized: hot and cold, wet and dry, or high sun and low sun. Europeans saw four divisions and called them winter, spring, summer, and autumn (or fall). Additional subdivisions recognized what we call the *cross-quarter days*. These cut the annual pie into eighths. Unlike the solstices, which are visible events in the middle and high latitudes, the equinoxes and cross-quarter days required greater degrees of observational sophistication.

The prime spokes of this "Wheel of the Year" are set at the solstices and equinoxes, solar positions easily measured today; their dates and times are precisely determined. Equinoxes occur when the sun stands directly over the equator. The summer and winter solstices (as defined in the northern hemisphere) mark the highest and lowest sun positions in the annual sky, when the sun is over the Tropics of Cancer and Capricorn, respectively.

Today, we tend to take these astronomical moments to signal starts for the spring, summer, autumn, and winter seasons. The ancients saw things differently. For them, the equinoxes and solstices were midseason rather than start dates. Using the sun for guidance, the summer solstice signaled the midpoint of the solar-dominated days, while the winter solstice pinpointed the central day of the night-dominated days, and festivals were celebrated accordingly. The spring equinox signals the day the sun springs above the equator in the north and daytime begins to exceed dark time; the autumnal equinox, the day the sun falls below the equator (from a northern perspective) and dark again dominates.

Of the remaining wheel's spokes, the cross-quarter days are lesser known today, at least in their original context. One, *Samhain* in the Celtic tradition and All Hallow's Eve in the early Christian traditions, is better known as one of North America's most celebrated non-holidays: Halloween. Another definitely gets a load of press each year. The Celts called it *Imbolc*, the Christian Church, *Candlemas*. We know it best as Groundhog Day. Lesser known is *Beltane* or May Day, and almost forgotten is *Lammas*, the first harvest festival celebrated at the beginning of August.

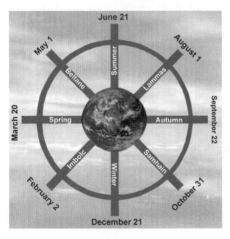

The Wheel of the Year shows seasonal and cross-quarter dates.

Seasons can be defined in myriad different ways depending on the ultimate application and need not all be of equal length. This concept has even moved into the non-science realm. For example, the baseball season is divided into off-season, spring training, regular season, and the playoffs or championship season. Then, the off-season returns, and the cycle begins over again, a rolling procession that repeats, more or less regularly, each year. Ice hockey follows the same four seasons (though spring training is called training camp), but they occur approximately in opposition to baseball's seasons.

Weather-related seasons are most commonly defined by North Americans in terms of temperature regimes: with summer the hot season, winter the cold, and spring and autumn transition periods. (On the Indian Subcontinent, two seasons, wet and dry—known as the monsoons—are recognized in the year, with abrupt rather than slow transitions.)

Another way of defining the seasons is by the *solar year*, as measured by length of daylight or potential daily solar energy received. When you divide the year into quarters and define the quarter with the most solar light/energy as *solar summer*, the least solar light/energy quarter as *solar winter*, and two transitional quarters of *spring*ing sunlight and *falling* sunlight, the midpoints of solar winter and solar summer would fall on the solstices, and the midpoints of solar spring and solar autumn would fall on the equinoxes. The start dates of these four solar seasons would occur about forty-five days ahead of those midpoints. These are the cross-quarter days.

Our Wheel of the Year now has eight spokes, each meeting the Gregorian Calendar at a specific date.

We start the year on 1 January, at no particularly interesting point of the solar year cycle, except perhaps, *perihelion*, Earth's closest approach to the sun, which occurs a few days later. It doesn't much matter where on a circle you begin. The ancient Egyptians started their year on the day that Sirius, the Dog Star and second brightest in the heavens, rose in the same sky-place as the sun. Cultures and traditions that follow a mostly solar calendar begin the year at differing points on the solar wheel. In some, the *spring equinox* begins the year. In others, it is Yule or the winter solstice that marks the start. The English originally started the year on 25 March, but moved it to 1 January around 1772. For a few other cultures, Samhain/All Hallow's Eve marks the transition. And those that follow lunar or lunar/solar calendars, such as the Chinese and Islamic traditions, have other starting dates that vary over the years.

Being a solar-driven person, I personally take Groundhog Day, the approximate start of solar spring, as the start of my year—the dark-day period has ended, and the daylight now begins a very rapid increase in length. For purposes of this book, however, I have chosen the winter solstice to start as it is close to the official New Year date and marks the day of shortest day length. This is a good place to begin a discussion of the weather, which is, after all, solar-powered.

WINTER

Three

21 December—The Winter Solstice

I breathe a sigh of relief today. It will be another year before I again hear the media decry the occurrence of wintry weather before this date, declaring it the "official" start of winter. But the winter solstice is official only in the minds of the media: Mother Nature and Father Sky have different ideas. I do laugh at statements by American and Canadian media that treat wintry weather in December as if it is abhorrent to have such conditions prior to the solstice. It makes me recall the phrase in the musical *Camelot*, where Arthur sings that winter is forbidden 'til December. In most of Canada and large portions of the United States, by 21 December, winter usually has become well established.

The day of the winter solstice is an astronomical marker rather than a meteorological or climatological one, "the day the sun stands still." The word "solstice," in fact, derives from the combination of *sol* meaning "sun" with *sistere* meaning "to stand still." It is the date when the sun's southward swing in the sky reaches its farthest point, lying over 23.5° south latitude, the Tropic of Capricorn, and bringing high summer to the residents of the southern hemisphere. (Santa Claus must get very warm in his heavy wool suit in Australia!) On this day, the solar journey across the heavens stops momentarily (from our relative vantage point on Earth) as it changes direction. Tomorrow, it will be marching back north toward the equator and the Tropic of Cancer.

This day will also have the year's shortest northern hemisphere daylight duration, though for many Arctic regions, the sun set for the long winter night as many as three months ago. For regions north of the Arctic Circle, the sun will not rise today at all; the best one can hope for is a twilight sky. Despite the shortness of daylight hours, this is neither the day with the earliest sunset nor latest sunrise. The earliest sunset occurred a couple of weeks back, in early

December; the latest sunrise lies ahead in early January, with both dates mildly dependent upon the latitude. These offsets result from the annual variation in the sun's *declination*—its angle above the horizon at solar noon—and factors related to Earth's rotation and orbital movement relative to the sun.

To account for these effects, timekeepers adjust our sunrise and sunset times using the *equation of time*. Determining sunrise and sunset times has two components. One accounts for the day-to-day changes in the inclination of the planet's equatorial plane to its orbital plane around the sun, the second accounts for our orbit being elliptical and not exactly circular. These factors actually make the day slightly longer than twenty-four hours from mid-November to early February, reaching about thirty extra seconds in late December. From early December to early January, the equation of time has its greatest impacts on sunrise and sunset times, overriding the changes in the solar declination.

In the weeks leading up to the winter solstice, the equation of time and declination work in opposite directions on sunset time. The declination pulls it earlier, while the equation of time pushes it later. By 8 December, the equation of time becomes dominant, and sunsets start occurring later. Meanwhile, both effects are pushing sunrise later and later. After the solstice, the situation reverses; both effects now combine to move sunset later in the day. However, while the declination now works to pull sunrise earlier, the equation of time continues to push it later. The equation of time prevails until about 5 January, the date of latest sunrise at 40° north latitude. Hereafter, the declination gains dominance, and sunrises begin to occur earlier.

The quarter-year period centered on the winter solstice marks the solar dark season that began back around Halloween. This day is Midwinter's Day—the contra point to Midsummer's Eve—and is the turning point from which the sun struggles back toward its dominance in the northern skies, a very slow pace to begin with, but one that accelerates until the spring equinox. At the North Pole, the sun will not return until March, but elsewhere the turn signals another start of the cycle, a new year, and a great place to begin our journey through everyday weather.

Four

Snowflakes: Winter's Crystal Lace

When does winter really begin? The day the total hours of daylight fall below a predetermined length or the temperature sinks below an established standard? The day of winter solstice? It's a hot topic for debate around the table or a warming fireplace. For me, the first snowfall that sticks on the ground for at least a day ushers in winter.

Snowflakes. Perhaps I should be more scientifically correct and call them *snow crystals*. Although many people call most frozen precipitation—other than hail or sleet—snowflakes, meteorologists technically use "snowflake" to refer to an assemblage of individual snow crystals that have collected together during their formation and descent. Snowflakes typically form when the near-surface air temperature is not far from the freezing mark. At this temperature, snow crystals are stickier, and those that collide adhere to one another better than at a colder temperature. At very cold surface air temperatures, the crystals do not stick together well at all. Thus, bitter-temperature snowfalls are mostly made up of snow crystals.

Snow Crystals

Snow crystals are made up of ice molecules and are typically 0.5 to 5.0 millimetres (0.02–0.20 in.) in size. Snowflakes are bigger—generally about 10 millimetres (0.39 in.) across and as large as 50 to 100 millimetres (2–4 in.) in diameter. Exceptionally large snowflakes have exceeded 200 millimetres (8 in.) and aggregate hundreds of individual crystals. For a snowflake to grow exceptionally large, however, conditions must be perfect. Besides needing temperatures near freezing that promote stickiness, large flakes usually form only when winds are very light—stronger winds break up large flakes as they fall. The biggest snowflake reportedly measured 380 millimetres (15 in.) across and fell on 28 January 1887 at Fort Keogh, Montana.

Another giant fell in Bratsk, Siberia, in 1971—a flake 200 by 300 millimetres (8 x 12 in.). Two decades earlier, residents of the English town of Berkhamsted saw snowflakes as big as saucers, almost 130 millimetres (5 in.) across.

We think of snow crystals as six-sided frilly stars and flat like a cookie. Snow crystal structure, however, takes on many different forms. Besides the six-armed stars that look like frozen lace, there are shapes ranging from flat hexagonal plates to small three-dimensional hexagonal pillars with hexagonal plates at each end, popularly called *cufflink crystals*. Other snow crystals resemble Dorian columns, some look like oak leaves, and some are shaped like hexagonal dinner plates. Generally, most perfectly formed crystals do not survive the fall to Earth. Branched crystals lose arms when buffeted by winds within or beneath the clouds, producing irregular and unsymmetrical shapes.

Snow Crystal Formation

For centuries, scientists puzzled over why snowflakes have six sides. In 1611, astronomer Johannes Kepler took time away from peering at the heavens to look into the small snowflakes of the earth. In his essay "On the Six-Cornered Snowflake," Kepler postulated that snowflakes were made up of globules of condensed and frozen moisture—their symmetry arising from geometrical efficiency because they were well-packed arrays of tiny spheres.

It took more than three centuries to fully correct Kepler's concept. We now believe the shape of a snow crystal is determined at the

Snow crystals form in a variety of shapes.
(NATIONAL OCEANIC AND ATMOSPHERIC ADMINISTRATION/U.S. DEPARTMENT OF COMMERCE, NOAA HISTORIC NWS COLLECTION, PHOTOS BY WILSON BENTLEY)

molecular level by the atoms composing the water molecule. Each molecule contains two atoms of hydrogen joined to one atom of oxygen, forming H_2O, a V-shaped molecule. In water, oxygen has a more powerful hold on the electrons that the two elements share in their water-making bond. As a result of oxygen's tighter grip, the oxygen atom side of the water molecule becomes slightly more negatively charged while the hydrogen side becomes slightly more positively charged. Since opposite charges attract, the positive hydrogen atom of one water molecule tends to stick to the oxygen atom of a neighboring molecule.

In liquid and vapor water phases, water molecules jostle and dance around their neighbors, forming few, if any, stable links between molecules. When the mass of water molecules loses enough heat energy—that is, as the water cools toward the freezing point—bonds (called *hydrogen bonds*) between water molecules form more easily and break less often. Upon freezing, everything seizes up. Now, each water molecule is surrounded by four others, to which it bonds. The oxygen atoms are arranged hexagonally in layers. New ice forming on the growing crystal's sides requires less energy than if it was added as a layer on the top or bottom faces. As a result, the side faces expand more quickly, and a flat crystal structure forms. This provides the underlying sixfold symmetry of the crystal lattice that grows to become the visible snow crystal or platelet of snow.

Snow crystals are sensitive to the conditions under which they form, particularly air temperature and excess relative humidity, expressed as the *supersaturation* level—both within and outside the clouds. These factors not only affect how the crystals grow but the basic shape they will take. The colder the air temperature, the sharper the ice crystal tips. At warmer temperatures, the ice crystals grow more slowly and smoothly, resulting in less-intricate shapes, such as more needles and plates. Thus, star-shaped crystals generally form in high-altitude clouds; needle, column, or plate crystals usually originate in middle-height clouds; and a wide variety of shapes grow in low clouds.

For example, thin hexagonal plate-like snow crystals form in air at 0 to −3.8°C (32 to 25°F). In colder air (−3.8 to −6.1°C [25 to 21°F]), needle shapes occur. Long, hollow hexagonal columns appear from −6.1 to −10.0°C (21 to 14°F). Flowerlike plates form at temperatures from −10.0 to −12.2°C (14 to 10°F). Six-pointed stars, called *dendrites*, appear from −12.2 to −16.1°C (10 to 3°F).

Wilson Bentley, the Snowflake Man of Jericho, Vermont

We owe many of our initial impressions of snowflakes/snow crystals to the work of a self-educated farmer from Jericho, Vermont. Wilson A. Bentley combined a microscope with a bellows camera to become the first person to photograph a single snow crystal, in 1885. Continuing this work until his death in 1931, Bentley captured more than five thousand snow crystals on film.

Bentley was a man who drew his livelihood mostly from his family's farm. Each winter, his passion for observing and photographing snow's delicate crystal structure became an obsession. His father and brother could not understand why he wasted his time "fussing with snowflakes," and neighbors considered him "a bit cracked." But his mother understood the delicate balance of artist and scientist in him and encouraged Bentley to follow his own path in life.

Bentley saw snow both as an artist and a scientist; he was a lover of beauty in many forms. He not only strove to photograph the most splendid of nature's handiwork, but also to exhibit his work whenever possible. The solitary photographer eventually became a lecture circuit performer, extolling the beauties of nature to audiences around New England. The belief that no two snowflakes are alike probably stems from his observation that every crystal formed a masterpiece of design, never to be repeated. The photographer became a local, then national, legend and is today known as Snowflake Bentley and the Snowflake Man. His 1931 book, *Snow Crystals*, co-authored with U.S. Weather Bureau meteorologist W. J. Humphreys, contains more than twenty-four hundred snow crystal images from his vast collection.

Three-dimensional snow crystals. (NATIONAL OCEANIC AND ATMOSPHERIC ADMINISTRATION/U.S. DEPARTMENT OF COMMERCE, NOAA HISTORIC NWS COLLECTION, PHOTOS BY WILSON BENTLEY)

Blizzard!

Most of the snow whitening the ground this morning fell lazily during the night, dry, fluffy snowflakes forming a duvet over the sleeping city of Guelph, Ontario. By morning, the municipal snowplows had cleared the major streets sufficiently so that I was able to extract my car from the driveway to make the 5-kilometre (3-mi.) trek to the office. The snow that fell overnight came mostly from the region ahead of the storm's cyclonic center, ahead of the warm front.

At first light, we did not know that the worst of the storm's punch was yet to come. The forecast warned of blizzard conditions, and a Winter Storm Warning was posted for much of southern Ontario, but like those venturing forth in the eye of a hurricane, we allowed the brief respite of this storm system's warm sector to lull us all into a false sense of security. We headed off to work and school with no thought of possible conditions in the hours ahead.

Just before noon, the storm's full fury was unleashed. The cold front rushed quickly over us and the temperature plummeted—first 10, then 20 C° below freezing (18 to 36 F°).[1] The wind, shifting from moderate southwesterly to gusty northerly gales in a heartbeat, whipped the newly fallen snow back into the air. Some of the airborne snow quickly crashed back down, only to be recaptured in the turbulent gusts and sent farther on its way.

Snowdrifts began to crest on the snowy sea, and, from them, ice crystals hissed as they ran across the snow surface like an army of

---

1 Note that I use "Celsius degrees" (C°) rather than "degrees Celsius" (°C). Strictly speaking, the former is a temperature difference and the latter a specific temperature on the scale. For example, 3 C° is the change in temperature of 3 degrees, while 3°C is a temperature just above the freezing mark. We generally use degrees Celsius for both in everyday usage, which is acceptable when sticking with one temperature scale. But in North America, with Canada using the Celsius scale and the United States the Fahrenheit scale, this can cause confusion. I have frequently seen people convert a 10 C° temperature change into a 50F° temperature change because they used the formula for converting temperature scales. The correct conversion is 10 C° = 18 F°, because 1.0 C° is 1.8 F°.

Young steer shivers through a blizzard in South Dakota. (NATIONAL OCEANIC AND ATMOSPHERIC ADMINISTRATION/U.S. DEPARTMENT OF COMMERCE, NOAA HISTORIC NWS COLLECTION)

beetles fleeing the light. Other snowflakes and ice crystals rose high into the sky on a journey to points unknown, far from their launch point. The massive, migrating population reduced the visibility to mere metres. Only when the wicked wind paused to recharge did we catch a glimpse of the buildings across the courtyard.

The quickly deteriorating conditions sent us home early. My usually short dash home became an odyssey as snowdrifts marched across streets and furious flurries of flying flakes made each inter-section a nerve-racking crossing of blind hope. After a forty-minute trip filled with detours, I arrived home to find my driveway drifted over with knee-deep snow—all that I had shoveled off it this morn-ing, plus new contributions from my neighbors' lawns and the curbside banks. But I, at last, was home and safe.

Whereas the previous night had been a winter wonderland of light snowflakes dancing in the wind, this night, a screaming ban-shee wind drove stinging, hard ice crystals against all that stood in its path. From inside, I heard the angry chatter of ice pellets racing across the attic roof. Around the corners of the house, the moaning wind sang a sorrowful blues. Blizzards are a time for rest, and who can argue with the wind.

Technically, *blizzards* are more than just heavy snowstorms. The snow that fell last night drowned the city in a sea of snow, but the mild temperatures and gentle breezes did not qualify the storm for blizzard status. The official Environment Canada definition for a blizzard is a winter condition meeting criteria for wind and visibility restrictions and regionally specified temperature and duration conditions. A blizzard must have winds exceeding 40 km/h (25 mph), with visibility reduced by falling or blowing snow to less than 1 kilometre (0.6 mi.) and must last for at least three hours. Technically, snow need not fall from the clouds; therefore, there are no accumulation criteria.

Regional Temperature and Duration Criteria for Blizzard Conditions

Region	Temperature at or below (°C / °F)		Minimum duration (hours)
Atlantic Provinces	−3.0	27.0	3
Quebec	−17.0	1.4	6
Ontario	−8.0	18.0	4
Manitoba/Saskatchewan	−10.0	14.0	6
Alberta	0	32.0	3
British Columbia (interior)	−10.0	14.0	6
British Columbia (coastal)	5.0	41.0	6
Northwest Territories, Yukon, and Nunavut	0	32.0	6

South of the Canadian border, true blizzard conditions occur, according to the U.S. National Weather Service, when winter storm wind speeds exceed 35 mph (56 km/h) and visibility is reduced by falling or blowing snow to 0.25 miles (400 metres) or less. In addition, these conditions must last for at least three hours. A severe blizzard boasts winds in excess of 45 mph (72 km/h). There is no specific temperature criterion in the American definition, but blizzards commonly have air temperatures below 20°F (−7°C) and severe blizzards below 10°F (−12°C).

Naming the Storm

The word "blizzard" first appears to have been used in the early 1800s, when it meant a severe blow, a cannon shot, or a volley of musket fire. Legendary American frontiersman Davy Crockett wrote of taking a blizzard (rifle shot) at a deer and speaking a blizzard (a strong retort or verbal blast) during a dinner speech. According to Canadian weather expert David Phillips, around this time period in the English Midlands a similar word, *blizzer*, referred to a severe storm of wind and snow.

American weather historian David Ludlum, in his two-volume work *Early American Winters*, cites an Iowa newspaper as the first published source of the word "blizzard" referring to severe winter weather conditions. It appeared in the 23 April 1870 issue of the

Where are the cars? The Victoria, British Columbia, blizzard of 1996 buried most vehicles.
(KEITH C. HEIDORN)

Estherville Vindicator. The paper's editor, defending a local citizen from remarks written in the rival *Upper Des Moines* of Algona, Iowa, wrote: "Campbell has had too much experience with northwestern 'blizards' [*sic*] to be caught in such a trap, in order to make sensational paragraphs for the *Upper Des Moines*."

The storm he referred to struck western Iowa from 14 to 17 March 1870. The editor of the *Upper Des Moines*, who also was the local Smithsonian weather observer, described the blizzard in his notes: "The storm continued with unabated fury (wind N.N.W.) all night and the following day . . . The air was so completely filled with drifting snow that no object could be seen at fifty feet distant. About ten inches of snow fell which is so drifted that the roads were completely blocked . . ." (Smithsonian Institution Records)

A week later, the *Vindicator* carried a story headlined "Man Frozen at Okoboji, Iowa." In it, "blizzard" is spelled with the accepted two z's: "Dr. Ballard, who has just returned from a visit to the unfortunate victim of the March 'blizzard', reports that his patient is rapidly improving."

Interestingly, in a 14 May story that year, the *Vindicator* reported on the reorganization of the local baseball team. Among the items of discussion was the renaming of the team from the Westerners to the Northern Blizzards. The reporter confessed to "a certain liking for it, because it is at once startling, curious and peculiarly suggestive of the furious and all victorious tempests which are experienced in this northwestern clime."

Thereafter, the term "blizzard," referring to a winter snowstorm, quickly became common usage to describe a severe winter snow-storm whose driving winds propel ice and snow through the air like a "volley of musket fire."

Regional Characteristics

Blizzards have different characteristics around North America. In the Arctic and northern Canadian prairies, blizzard snowfalls are generally light with visibility reduced by blowing snow under gale-force winds. On the northern Pacific coast and in its coastal mountains, blizzards result when strong Pacific storms drive relatively warm, moist air against extremely cold air masses pushing out of the interior plains and valleys through the mountain passes.

Blizzards forming east of the Rockies, ushered by *Alberta Clippers* and *Colorado Lows*, channel severe cold outbreaks from arctic latitudes to as far south as Texas, where they are known as *Blue Northers* (see page 239 to read more about Alberta Clippers and Colorado Lows. Snowfall may be very heavy in southern regions, where arctic air combats tropical Gulf of Mexico air. Such storms may paralyze the Missouri, Mississippi, and Ohio River valleys, as well as the Great Lakes and Atlantic coast.

Colorado Lows may also move over the southern tier of American states only to gain strength as they move offshore over the Gulf Stream's warm waters. Newly reborn, these storms are called *Hatteras Lows* or *Nor'easters* when they bring blizzard conditions to the Atlantic coast from the Carolinas through New England to the Canadian Maritimes and Newfoundland.

One unique form of blizzard strikes the southern and eastern shores of the Great Lakes. This brand of storm develops to the rear of intense low-pressure systems and may persist for several days after the low has moved away, feeding off the cold air surging south. These *lake-effect blizzards* blow snow squalls onto the lee shores of the Great Lakes as the intensely cold arctic air mass picks up moisture while crossing the relatively warm lake waters. Where and when the snow is not falling, the arctic gales whip ground snow into the air, thus reducing visibility to near zero.

In the blizzard that opened this chapter, the first stormy day resulted directly from an approaching intense low-pressure system crossing over the lower Great Lakes. In its wake, a strong persistent flow of cold air became established on the eastern flank of a large arctic air mass, which extended from northwestern Ontario through Illinois. Lake squalls, moving off the water of Lake Huron and Georgian Bay, waxed and waned for several days before finally diminishing.

As the winds howled outside and snow and ice crystals danced

Man in blizzard. (ARTWORK BY KEITH C. HEIDORN)

vigorously across the sky, I snuggled deeper into my cozy chair, sipping hot chocolate. Enjoying the winter's rest, I wrote and napped and read. As I was about to doze off once again, my eyes caught the words of Henry David Thoreau, writing from Concord in 1853: "All day a driving snowstorm, imprisoning most, stopping the cars, blocking up the roads. No school to-day. I cannot see a house fifty rods off from my window through [it]; yet in midst of all I see a bird, probably a tree sparrow, partly blown, partly flying over the house to alight in a field. The snow penetrates through the smallest crevices under doors and side of windows." (*Journal*, 1853)

I too was happily imprisoned—and enjoying every minute.

Snowdrifting

Dry snow fell in copious amounts all through the night. The accumulation was daunting enough, but that it fell like the emptying of a million feather pillows into a hurricane only increased the misery. We could have coped, at least with some degree of sanity, with the 20 centimetres (8 in.) of snow over the community, but having it whipped into waist-level—and higher—drifts across traffic lanes and driveways elicited words of displeasure not suitable for sensitive ears.

The snow, blowing amid gale winds during the early morning hours, snarled rush hour traffic. Of course—to paraphrase Will Rogers—it was impossible to call it traffic as no vehicle was moving faster than a snail's pace, and the only morning rush was that of the wind, or that exhilarating feeling of trying to walk against the blow. Those walking with the wind, however, were rushed forward. I had chosen foot travel, so at least I was moving, though not as fast as the scurry of snow along the whitened snowbanks lining the sidewalk.

Just before noon, the storm's full fury was unleashed. As the cold front rushed quickly over us, the temperature plummeted and the wind shifted to gusty northerly gales, whipping the newly fallen snow high into the air. Some airborne snow quickly rushed back down, only to be dispersed by gusting wind. Snowdrifts began to form snow dunes.

The blowing, and subsequent drifting, of snow can change a snowstorm or post-storm event into a blizzard even if no new snow falls from the clouds. The blowing snow, like blowing sand and dust, creates its first hazards by reducing visibility. But, to quote the old saying, what goes up, must come down, and generally a good episode of blowing snow builds and sculpts snow dunes worthy of any sandy beach or desert landscape. Heavy snow accumulations can prove a major obstacle to humans, but rearranging it into drifts metres high produces a nightmare in white fluff.

The Physics of Blowing and Drifting Snow

Snow becomes airborne when wind catches falling snow or lifts loose snow crystals from the underlying surface. The process is in some ways analogous to blowing dust and sand, except that snow is much less dense and can change phases (melting, sublimating, refreezing) under normal environmental conditions. Falling snow may become windblown by even the lightest winds. For the wind to lift snow from the surface and carry it, its speed must exceed 13 km/h (8 mph). Once lofted, snow crystals will remain in suspension until the wind speed drops or the crystals enter a region where surface features reduce the wind speed.

The quantity of snow carried by the wind, and the distance it travels, depends not only on the wind speed and its level of gustiness (*turbulence*), but also on air temperature, the nature of the snow itself, and the fetch over which the wind is blowing. Light, dry snow, formed under cold air temperatures, is ideal for blowing and drifting, while heavy, wet snow takes more wind force, and icy snow surfaces resist even the strongest winds. A gust may raise a small flurry of snow from a rooftop, but run a strong wind across a flat field, and the amount of raised snow can be prodigious, measured in tonnes (tons).

Once airborne, individual snow crystals dance and cavort with the wind as it sweeps across a landscape and among, beneath, over, around, and even through terrain features. On the smallest scale, called the *microscale*, the wind accelerates and decelerates in response to the presence of features as small as bush or plant stems and as large as mountains and skyscrapers. For example, a relatively exposed hilltop bare of tree or bush, even one only 1 to 2 metres (4–7 ft.) above the surrounding terrain, can be swept clean of snow, and in the absence of other obstacles, that snow may be deposited on the hill's lee (downwind) side or in some downwind depression in the surface. Then again, it may land in the next county.

On very flat terrain, such as a frozen lake surface, tundra, or prairie, snow can blow for tens, even hundreds, of kilometres

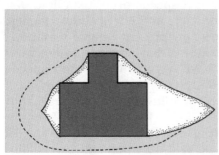

Snow-drifting pattern around a building (grey mass). Dashed line indicates uniform snow location without wind.

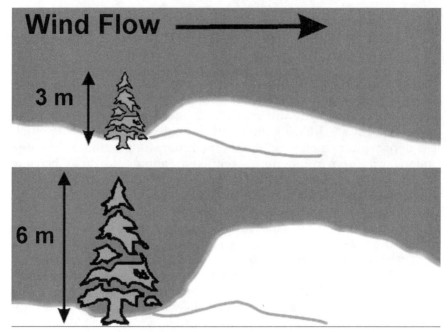

The size of an obstacle, here a tree, affects the resulting drift size and shape.

before being redeposited on the ground or among buildings or vegetation. In such terrain, drifts may form in shapes similar to sand dunes. For example, *transverse sand dunes*—large dune fields consisting of ridges of sand with a steep face on the downwind side form where sand is abundant and wind flow is constant in direction—have a snow equivalent in the Arctic called *sastrugi.*

In terrain where obstacles of many types, sizes, and shapes are present, snowdrifts can form a very complex "dune" field. Obstacles that present a barrier to unobstructed wind flow form a drift in the region of the turbulent flow when wind speeds drop, usually a distance behind the object. Obstacles such as hedges, trees, fences, buildings, and even piles of plowed snow present unique variations in the wind flow pattern. Not only do they have regions where wind speed diminishes, they often also have regions where wind speeds accelerate. Such acceleration zones are void of drifts and may have their snow scoured to the ground, becoming the source region for downwind drifts.

The height and breadth of the obstacle affects the nature of snowdrifting around it as does its *porosity* (the percentage and size of "holes" in the barrier). As a rule of thumb, drifts extend downwind from an obstacle of moderate height to six to nine times the object's height.

Snowdrifts form before a townhouse door in Guelph, Ontario. The snow-free zone close to the building results from the wind flow pattern. (KEITH C. HEIDORN)

Preventing and Reducing the Impacts of Snowdrifting

Snowdrifting can be more than just a nuisance to snow-removal efforts. When snow drifts around buildings and other structures, it can cause safety hazards and even trap occupants within, blocking doors and exhaust ports and building up on roofs, which can lead to a collapse. To reduce and prevent such hazardous snow accumulations, scientists, engineers, and architects have sought to understand the nature of snow accumulation. Once understood, mitigation measures can be proposed and assessed.

I was once an applied meteorology/climatology consultant for an engineering firm in Ontario. One of its services is the study of *snow loading* and snowdrifting. (Snow loading is the weight that can be carried by a structure, such as a flat roof, when snow is piled on it.) Their projects included assessing the impacts of snowdrifting around buildings, drifting and loading on roofs, and drifting along roads and railways. To study snowdrifting, they use a water flume into which a specific sand type is dropped into the moving water.

When the sand is sifted over the water surface like falling snow, the water currents carry it in a manner that simulates snowfall and snowdrifting on a smaller scale. By placing a model of a building, a building complex, or section of roadway terrain in the water flume, the engineers can simulate how snow will drift around the full-sized configuration under a specific wind direction. Usually several directions are tested that corresponded to climatological storm-wind or high-wind directions. When the waterflow is stopped, the height and locations of the sand piles are noted, and if the test shows a possible problem area, a remediation can be proposed and tested. This may involve the adding or subtracting of a building element, landscaping, a fence or other feature to the configuration.

One example of this work concerned a university library where drifting snow would often block emergency exit doors. By adding a small element to a building wall similar to a windbreak, the drift was moved several metres out from the building instead of forming against the exits. By using such simulations, engineers and architects are not only able to correct problems after the fact but also able to test designs before construction. The preconstruction remedy may be as simple as a small change in landscaping or in the orientation of a building on the property. For example, placing the long axis of a building in the direction of the prevailing winter winds minimizes some problems arising from drifting compared to placing that axis perpendicular to the prevailing wind.

Similar expertise developed from observational field experience has been used to prevent or reduce snowdrift formation over rural roads through the use of snow fencing. In such cases, the fencing is

A March blizzard nearly buried the utility poles in Jamestown, North Dakota. (NATIONAL OCEANIC AND ATMOSPHERIC ADMINISTRATION/U.S. DEPARTMENT OF COMMERCE, NOAA HISTORIC NWS COLLECTION)

installed each autumn and removed in the spring. In other cases, landscaping, such as the planting of a row of low trees or hedges, can be used as a permanent snow fence with the same winter effect. The function of a snow fence is to cause wind-blown snow to settle before it reaches the site requiring protection. The fence must be erected perpendicular to the prevailing wind direction so the snow deposits in front of and behind it. This is termed *collector fencing* because its purpose is to collect the snow near the fence. Fencing may also be used to keep specific areas clear of snow by increasing the wind speed and blowing the snow elsewhere. Such fencing is called *blower fencing* or *deflector fencing*, depending upon its configuration.

Natural materials such as hedges—low coniferous trees and bushes—and earthen dams can be used in the same manner as snow fencing. Even properly placed and sized snow dams can be employed to reduce adverse drifting potential. Hedging has the advantage of being low maintenance but it requires time to grow before becoming effective. Natural snow-control landscaping has been used by farmers for years, as they learned by experience how best to corral snow.

Seven

Cold Wave: The North Wind Doth Blow

"**Upon retiring for the night**, I tried to blow out the candle," a Winnipeg, Manitoba, settler wrote in his diary one cold winter night in 1879, "but the flame was frozen, so I had to break it off."

Although the settler spun the cold truth a little loosely, the depth of January cold is famous across North America. That is not to say that North America has a monopoly on cold weather in the northern hemisphere. The very name Siberia conjures images of frigid cold, ice, snow, and perpetual blizzards. But North America is unique among the Arctic continents because its major mountain chains run north–south rather than east–west as in Europe (the Alps) and Asia (the Himalayas). This makes nearly all of North America susceptible to vigorous cold waves. Without the mountain barriers to block them, arctic cold outbreaks can sweep south toward the subtropical Gulf of Mexico, even crossing its waters on occasion.

The birthplace for most frigid outbreaks is the continental expanse of northern Canada and Alaska near or within the Arctic Circle. Here, long, dark winter nights couple with clear skies and a surface covered with snow and ice to progressively chill the air. Since the sun here does not rise or only skirts along the horizon for much of January, it provides no warming heat. Surface snow and ice reflect away what little sun weakly beams down. To compound the lack of incoming heat, snow very effectively radiates away what little heat it has, thus dropping the surface air temperature further, until it plunges to the temperature of the high atmosphere, many tens of degrees Celsius below freezing.

As the air chills, it also becomes denser, forming large domes of high surface pressure. Eventually, this cold dome breaks its bond with the spawning ground and rushes wildly southward. Howling winds precede the great air mass, announcing its advent. Trees shudder. Birds shiver. Rabbits burrow deep within snowbanks. As many a Texas Panhandle farmer has worried, "The only thing standing between us and the North Pole is a three-strand barb wire fence. And I fear one strand is missing."

Long before they understood the scientific and geographic elements behind the weather, inhabitants of high northern latitudes heard the voices of gods and demons in the howling rush of frigid air leaving the polar region. The Greeks saw the North Wind as a gray-haired old man, strong in body and harsh in disposition, named Boreas. Boreas has many relatives: the twin brothers Norther of the Texas Panhandle and Blue Norther of Alberta; the American blood brothers Arctic Screamer, Stikine, and Blizzard; the Asian "Steppe-sisters," Steppenwind and Buran; an uncle from Greenland, Nekrayak; the Siberian cousins Viuga, Purgas, and Myatel; and the Scottish nephews Blaast and Landlash.

North America generally has the fiercest frigid air outbreaks, but across the northern hemisphere, the home of winter cold is Siberia. Two locations in Siberia currently hold the hemispheric record for coldest reading and the coldest global surface temperatures outside the Antarctic. The first was measured on 7 February 1892 in Verkhoyansk, the second on 6 February 1933 at Oimekon, both a numbing $-76.8°C$ ($-106.2°F$). The large expanse of the Eurasian continent and the east–west mountain ranges contribute to producing such extreme temperatures. Together, they block the passage of warmer moist air into Siberia and hinder the passage out of cold air masses. As a result, the low humidity allows surface radiation to stream out into space, and Siberian air masses grow colder and colder during the long winter nights. However, you can't keep a cold high bottled up forever; eventually, it becomes so massive it squeezes out of its natal ground, sending cold air blasts to the outside world, even across the Arctic Ocean into North America.

I have never experienced a Siberian cold blast in its native land—watching *Doctor Zhivago* was enough for me—but I have experienced Arctic Screamers and Barbers that originated in the Siberian deep freeze descending across the Great Lakes region. In North America, we term such cold outbreaks the *Siberian Pipeline*, and they can bring hazardous conditions across most of Canada and the northern United States, particularly the continent's northeastern quadrant.

Among my most lasting memories of extreme cold is the severe winter of 1976–77. Temperatures in southwestern Ontario in January dropped to $-30°C$ ($-22°F$), ushered in on winds howling over the white landscape at speeds of 40 km/h (25 mph), with gusts reaching 70 km/h (44 mph) at times. As the wind raged across the dry snow, blanketing the countryside, it caused blizzard conditions and whiteouts for days on end.

In 2004, a Siberian Pipeline established itself across the subarctic into northeastern North America. Bone-chilling cold descended, setting or approaching all-time record cold at many locations. The cold flooded south on strong winds that drove windchill conditions to deadly levels. At the Mount Washington Observatory in New Hampshire, the summit weather station recorded a gust of windchill below −73.3°C (−100.0°F). Such cold is extremely hazardous to urban residents not accustomed to deep and long-lasting cold spells. But the coldest air temperatures experienced in New England, Quebec, and the Maritimes (in the −40°C [−40°F] range) were still far from the continental record set nearly six decades ago in Canada's Yukon.

The record books generally give it one line: "Coldest Temperature (North America): −63°C/−81.4°F, Snag, Yukon, Canada, 3 February 1947."

The Brutal Winter Season of 1946–47

Snag is located in the Yukon's southwestern corner and was, in 1947, the site of an auxiliary military airfield, about 6 kilometres (4 mi.) south of the village. The airfield had been built during the Second World War as part of the Northwest Staging Route to provide weather observations and an emergency landing strip for the Royal Canadian Air Force, the U.S. Army Air Corps, and civilian air traffic heading to the much-forgotten Northern Pacific Front.

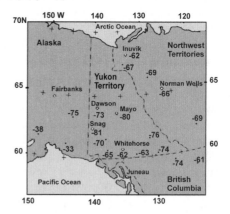

Extreme low temperatures (°F) recorded during North America's great cold period in January and February 1947.

In 1947, Snag boasted a population of eight to ten Native people and fur traders, with an additional fifteen to twenty airport personnel—meteorologists, radio operators, and aircraft maintenance men—who lived at the airport barracks. Snag sits in a bowl-shaped valley along the White River, a tributary of the mighty Yukon River, which flows northward from the Wrangell Mountains near Mount Churchill. The village is about 30 kilometres (19 mi.) east of the Alaska border and 25 kilometres (15 mi.) north of Alaska Highway Mile 1,178. The St. Elias Mountains rise 50 kilometres (30 mi.) to the south. Local vegetation on the unglaciated uplands is mostly scrub and poplar trees less than 6 metres (20 ft.) tall.

The Snag region has a unique climate. With the Wrangell, St. Elias, and Chugach mountain ranges forming a high barrier along the coast, the flow of milder Pacific air into the region is effectively blocked. This leaves the region open to the western Arctic's cold frigid air masses. Additionally, cold air from the coastal mountains' north slopes drains down the broad valley terrain surrounding Snag and accumulates in low areas. The cold air drainage under light winds forms ideal conditions for extreme low temperatures.

Snag's climate in winter thus resembles eastern Siberia. Winters are prolonged, characterized by bitterly cold temperatures, generally clear skies, and calm or light winds. January in Snag sees the average daily maximum temperature at a chilling –25.1°C (–13.2°F) and the average daily minimum temperature at a frigid –35.6°C (–32.1°F); the thermometer regularly dips below –50°C (–58°F).

North America's cold extreme was not an isolated event in either time or space. Extreme cold conditions were the norm over Alaska and northwestern Canada for much of the winter of 1946–47. A new low-temperature record for the Yukon was set on 13 December 1946 at Mayo, 260 kilometres (163 mi.) northeast of Snag, when a numbing –72°F (–57.8°C) was recorded.[2]

Meteorologists attributed the extended cold weather to a strong zonal (westerly) circulation in the upper atmosphere across North America that trapped arctic air over Alaska and northwestern Canada for much of the winter. This built a dome of intensely frigid air over the Yukon. The air mass, which spilled into northern British Columbia, the Northwest Territories, and eastern Alaska, reached its coldest between about 27 January and 4 February. Many sites set record cold temperatures for January and February that still stand. For example, British Columbia's all-time officially recorded low temperature descended on Smith River on 31 January: –74°F (–58.9°C).

As the cold, heavy air mass lingered over the Yukon, its coldest air settled near the surface and drained into low-lying valleys. As February dawned, weather observers at Snag reported clear skies and calm winds—ideal conditions for extreme cold. Overnight temperatures nearly slid off scale and during the day rose no higher than –50°F (–45.6°C).

Few ventured forth that Groundhog Day (2 February) to look for shadows as the morning temperature dropped to the lowest mark on the standard alcohol minimum thermometer: –80°F (–62.2°C), a new record low for the continent.[3] By 2:00 PM, the temperature had risen to only –51°F (–46.1°C). Shortly after the day's temperature peak, the

---

2 At the time, official temperature records in Canada and the United States were kept in degrees Fahrenheit.

3 In regions where extreme cold temperatures may occur, thermometers are alcohol-based rather than mercury-filled, as mercury freezes at –39°C/F.

alcohol plummeted rapidly over the next five hours, then declined more slowly through the night.

Weather officer-in-charge Gordon M. Toole kept vigil on the minimum thermometer that night. Shortly after 7:00 AM on the morning of 3 February, Toole cautiously opened the instrument shelter door, taking care not to breathe on the instruments inside. Using his flashlight to illuminate the minimum thermometer, he saw that the sliding scale within the alcohol column, used to register the minimum temperature, was below the lowest scale marking on the instrument: −80°F (−62.2°C). Using a set of dividers to determine the slide's position below that mark, Toole estimated the reading at about −83°F (−63.9°C). The previous record low had lasted only one day!

On that record-setting day, the morning observation reported a surface pressure of 103.7 kilopascals (1,037 mb), calm winds, and a visibility of 32 kilometres (20 mi.). In some directions, visibility was reduced by patches of ice fog. The snow on the ground measured 38 centimetres (15 in.) but, due to the air's intense dryness, was decreasing at a rate of about 1.3 centimetres (0.5 in.) per day. The record low was recorded at 7:20 AM, an hour and twenty-two minutes before sunrise. The high for the day reached only −56°F (−48.9°C).

When news of the record cold reached the media, it made headlines around the world. The *Toronto Globe and Mail* declared, "Snag snug as mercury sags to a record −82.6°[F]." Reporters from around the world sought interviews from the "frozen few" on the experience of life at minus eighty.

The third of February was the bottom of the cold air barrel for much of the Yukon as temperatures moderated over the next few days. The westerly air current that had locked the region in an icy prison finally relaxed its grip and pushed the cold air dome out. Released from its cradle, the cold slid southeast across the continent, bringing frost to northern Florida. Pacific maritime air replaced the arctic air and, for a few days, conditions became relatively balmy.

Since that day in 1947, several cold spells have descended on North America, but none have officially broken Snag's record. Prospect Creek, Alaska, came closest, falling to −80°F (−62.2°C) on 23 January 1971, setting the American record. (However, a −70°F [−56.7°C] reading at Rogers Pass, Montana, on 20 January 1954 remains the record for the lower forty-eight states. Only northeastern Siberia, Greenland, and Antarctica have recorded lower official mean temperatures than Snag.)

Two locations used in a permafrost study in a mountain valley near Fort Nelson, British Columbia, however, reached unofficial lows of −71.1°C (−96°F) and −68.9°C (−92°F) on 7 January 1982. The official

Urban canyon effects and blizzard conditions hamper a pedestrian in Rochester, Minnesota.
(NATIONAL OCEANIC AND ATMOSPHERIC ADMINISTRATION/U.S. DEPARTMENT OF COMMERCE,
NOAA HISTORIC NWS COLLECTION)

low recorded that day at Fort Nelson was a relatively warm –42°C (–43.6°F).

Given the many pocket valleys in northwestern North America, it is not surprising that the right combination of conditions could produce a spot temperature lower than –80°F (–62.2°C). Perhaps it is just a matter of time before an official climate station resides in one. Snag, after all, only had a weather station and an airport to support the war effort. An automatic weather station might be required in the future in some remote valley to monitor precipitation for hydroelectric power generation, or assess wildlife conditions. While performing its required task, the monitor just might be in the right place at the right time. After all, records are made to be broken.

Eight

Too Cold to Snow?

During the depths of January cold, we often hear some self-proclaimed weather sage observe "It's too cold to snow." At first, the statement appears to jibe with general experience. Usually, the coldest days and nights are clear and dry. While this is generally true—large, frigid arctic air masses are typically characterized by dry, mostly clear weather—it is not the cold that prevents snow falling in these air masses, but the lack of ascending air to start the precipitation process.

The truth is, it is never too cold to snow—but, it is also true that the colder the air temperature, the less water vapor the air may contain. In fact, the amount of water vapor at saturation drops by about half for every 10 C° (18 F°) drop in air temperature. So one of the factors that often makes very cold air snowless, or seemingly so, is its low moisture content. Even if all the moisture were to drop out as snow, it might not make a bowl of flakes.

We can go deeper, well, actually higher, to find another take on this issue. We live within a very shallow layer of the atmosphere, generally less than 100 metres (328 ft.) above Earth's surface, unless you have an apartment in a high-rise like Chicago's Sears Tower. The weather works its ways through a greater depth, mostly residing within the troposphere, the lowest 20 kilometres (65,000 ft.) of atmosphere. Through that depth, conditions are not uniform, however; in fact, above the boundary layer, the lowest kilometre or so, the temperature generally decreases with height but may contain several layers of warmer or cooler temperature relative to that immediately above or below.

Three of the general conditions needed for snow (or rain) are: the atmosphere must have sufficient moisture, the moist air must be lifted to a height where it can condense into liquid or deposit directly to ice, and the process must occur in an environment whereby the

A crisp winter day in southern Ontario brings bright sun and cold temperatures.
(KEITH C. HEIDORN)

cloud droplets and ice crystals can grow to a size large enough to survive a fall to the surface.

Those big, cold high-pressure cells that contain the coldest air usually do not form clouds because their dense, cold air sinks rather than being lifted, contrary to the second condition. Despite the cells' very cold nature, the sinking air is also warming a bit as it descends, taking conditions in the wrong direction for producing clouds. These conditions are, in themselves, sufficient to prevent cloud formation and snowfall from occurring within the air mass, but they are dynamic atmospheric processes.

At very cold temperatures, –40°C (–40°F) and lower, snow can actually fall out of the cleanest, clearest blue sky without intervening clouds if the air mass is at or below its frost point. Temperatures need not be so cold if dust or other minute particles are present in the air on which the water vapor may deposit. When *condensation nuclei* are present, *diamond dust* may form at temperatures just below –20°C (–4°F). At such temperatures, the water vapor in the air spontaneously forms ice crystals, which slowly settle earthward. When these falling ice crystals are caught in the light, they sparkle like gemstones, hence the name diamond dust. Usually a very light snowfall or snow shower that rarely accumulates to any appreciable amount, diamond dust may be observed as individual crystals on surfaces such as automobile exteriors and glass when they are cold.

Nine

January Thaw

Snow melts
Ice thins
Early buds tease of coming Spring.
—Keith C. Heidorn

I feel a bit odd writing about the *January Thaw* in a region of Canada that often has little to thaw in mid-January or at *any* time during the winter. Although my body now resides in "tropical" Victoria, my mind and spirit still wander across the Great Lakes region of my birth.

The January Thaw is a period of mild weather that supposedly recurs each year in New England and other parts of the northeastern United States. It holds a place in North American weather lore nearly as prominent as autumn's Indian summer. Tradition indicates the thaw should occur during the third week of January across the Great Lakes, St. Lawrence Valley, New England, and the Maritimes. As far as I can determine, it is unique to this continent, particularly in the aforementioned regions, approximately east of the Mississippi River and between 40° and 50° north latitude.

Before I get into the debate over whether this weather spell is as real as we imagine, I need to define *thaw*, which is an unusual extended period, over several days, of above-freezing weather that significantly reduces snow or ice cover. Another relevant term not likely heard outside professional circles is *singularity*, which is a characteristic meteorological condition that tends to occur on or near a specific calendar date more frequently than pure chance would indicate.

If we assume there is a January Thaw, how do we technically define it so that we know what to look for in the data? Things get a bit fuzzy because there are no fully accepted set of weather parameters to define it. Consensus holds that the January Thaw should

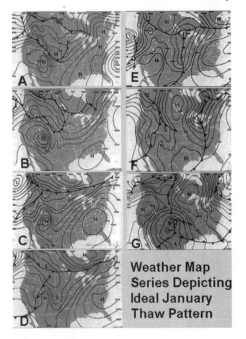

Weather Map Series Depicting Ideal January Thaw Pattern

Sequence of weather maps showing the typical development of a January thaw pattern over eastern North America.

last for several consecutive days, have maximum daily temperatures above freezing, and have mean daily temperatures that rise around 5 C° (9 F°) above the expected mid-January climatic average for a given location or area.

January Thaws usually follow a strong cold snap, but one need not occur every year. During the very cold winters of the late 1970s, there were no January Thaws in southern Ontario, eastern Canada, and the New England region during 1978 and 1979. In several winters during the 1990s, the thaw was absent or hardly noticeable because winter conditions were conspicuously mild. From historical records, a January Thaw is common in the northeastern regions of the United States and eastern Canada in at least two of every three years.

In prime January Thaw country, the ideal weather pattern characteristic of the thaw period unfolds in this manner, as shown on the accompanying weather maps. It begins after a very cold air mass from northern or western regions has slid over the region (panel A), eventually moving out over the Atlantic Ocean. As that cold air mass leaves, the Bermuda High strengthens (panels B, C, and D) and becomes positioned over the southern Atlantic coast or southeastern United States while a broad low-pressure trough moves slowly across northern Ontario and Quebec (panels B through E). The juxtaposition of the isobar patterns of these two map features (panel E) places the northeastern United States and southeastern Canada region into a south-southwesterly flow of warm air from the Gulf of Mexico. This air advances over the northern snow and ice fields and begins a thawing. Often during this time, the upper air wind patterns are in a period of readjustment, thus surface weather systems stall or creep slowly across the eastern continent's midlatitude belt. After several days of warmth, the regional weather again comes under the influence of a strengthened polar high (panels F and G), and cold weather returns.

All meteorologists would agree that a period of thawing is a common winter weather event in the continent's midlatitudes. Why then should such a thaw have a preference for the 20 to 26 January period rather than be a purely random event?

Whether there is a preferential calendar slot still remains the subject of debate among climatologists. Independent analyses by David Phillips of Environment Canada and Art DeGaetano of the U.S. Northeast Regional Climate Center at Cornell University show definite warm spikes in the climate records for many sites across the prime eastern region during this preferred period.

The other side of the debate is characterized by the studies of meteorologists Christopher M. Godfrey, Daniel S. Wilks, and David M. Schultz, who find no unique January Thaw period that is statistically significant. They see thaws during this month as purely random events. While they readily admit their analysis is not conclusive proof that no January Thaw exists, they note that no dynamic or plausible physical mechanism has yet been advanced to explain why a winter thaw should favor any specific narrow period year to year.

Godfrey and his colleagues do suggest one explanation for folklore and popular perceptions of weather singularities such as the January Thaw: human nature. Psychologists have long recognized a human tendency to want to find patterns in the world, whether or not those patterns actually exist. Seeing patterns in the environment, real or not, appears to have served our species, and many individuals, well in the past and will continue to do so. Better we see the face of a non-existent tiger in the bushes and be wary than not see the face of a real tiger and be killed.

That we see a pattern in the winter cycling of weather across certain regions that gives rise to a welcomed respite from the cold is a given. If it were not there, scientists would not have spent more than a century trying to prove or disprove its existence.

Is the January Thaw real? I have to emphatically say yes—not as a meteorologist or climatologist with proof irrefutable, but as a human who has seen its pattern unfold over decades. At times, I have cursed its coming when it melted away a good snow base for cross-country skiing. More often, I have welcomed its touch. I do miss looking forward to a January Thaw, living now in a land where snow cover is rare.

Solar Spring: Weathering Groundhog Day

The second of February is a rather inconspicuous date that would probably be ignored were it not for the media hype over a furry North American rodent called *Marmota monax*. You likely know it better as the woodchuck or groundhog, a character usually going by the name of Punxsutawney Phil or Wiarton Willie.

In the traditional Christian calendar, 2 February was known as Candlemas Day, a feast day commemorating the presentation of Christ in the temple through the bringing of candles to the church to be blessed by the priest. Earlier, in the British Isles, the important day was called *Imbolc*, one of the four cross-quarter days, an ancient Celtic festival honoring the Earth Mother; it was considered the first day of spring. Other cultures have traditions similar to the Celts regarding this day. As I mentioned earlier, this is the start of solar spring.

As 2 February arrives, the now-rapidly increasing day lengths in the temperate zones and increased solar strength in all but the most polar zones of North America begin to stimulate the internal clocks in many plant and animal species that will result in the bursting forth of new life in the coming months. Folk legends have long endowed certain animals with the ability to foresee the weather. The hedgehog, bear, and badger in Europe begin to stir from their hibernation around early February, and folk myths held the belief that they poked their noses out of their burrows or dens to see whether it was time to come out or time to roll over for another month. North America has no resident hedgehogs, so European settlers endowed the New World groundhog, or woodchuck, with this predictive insight.

43

In case you have been living in a burrow yourself, the legend goes like this: If the groundhog sees its shadow when it crawls out of its hole, it goes back in, indicating winter will continue for another six weeks. If it sees no shadow, spring is on the doorstep.

Is there any validity in the groundhogs' early forays? Not really. Of course, the larger an area over which a prediction is made, the worse it can be. For example, when Canada's Wiarton Willie makes his pronouncement on the advent of spring, much of Canada would be very happy to see only six more weeks of winter (which takes us to the spring equinox). In southern regions of the United States, spring is already under way and marching northward when Pennsylvania's Punxsutawney Phil is dragged from his burrow.

At times, the media goes (hedge)hog wild when the 2 February prognostication comes true and ballyhoos the furry marmot's intuition. When the prediction goes awry, Phil and Willie join the fraternity of professional weather forecasters whose forecasts have been wrong in the dog (or is it hog?) house.

We can, however, drag a sliver of truth from the legend. If the groundhog sees his shadow in the morning air, it must be clear. This may indicate the start of a period, perhaps only short, of sunny but cold days under a high-pressure system. If the shadow is not there because the sun is hidden by cloud, a nearby low-pressure cell or warm front might be bringing warmer, more spring-like air to its northern home, albeit briefly.

Eleven

Of Sun Pillars

In most of the Canadian interior, February signifies the turning of the corner on winter's extreme cold. Still, large arctic air masses do slip out of the Far North, bringing days of extreme cold to southerly regions of the country and into the United States.

I recall one such February arctic outbreak many years ago. The temperature for the day had barely exceeded −18°C (0°F) in the Guelph, Ontario, region, centered under a large arctic air mass. As I walked home from the bus stop in the late afternoon, the soon-to-be-setting sun illuminated the horizon in brilliant light. Looking beyond my parka hood, I noticed that, although the sky was clear, a soft fall of snow encircled me. The snow, so fine I could see individual crystals alight on my dark sleeve, floated around me, fluttering ever so slowly downward. In the low sunlight, the snow crystals glittered like diamonds against the darkening sky. The brilliance of the small illuminated crystals gives such snowfall its name: diamond dust.

As I continued my walk homeward, the sun sank lower in the sky and was only minutes away from setting. Looking toward the solar orb, I saw a crimson pillar of light rising from the top of the solar disk. This was a *sun pillar*, a feather of light that extends vertically above and/or below the sun, and the most commonly seen form of *light pillar*.

Most sun pillars are visible when the sun is low on the horizon (generally no more than six degrees above) or just below it. They usually extend only five to ten degrees of arc directly above the solar disk, though on rare occasions they are seen as high as twenty degrees above the solar disk.

Moonlight and strong artificial light, such as streetlights, may also generate pillars. *Moon pillars* form in the same fashion as sun pillars. Pillars arising from artificial lights can be greater in extent due to the nearness of the light source to the observer.

Pillars appear when snow or ice crystals reflect light forward from the strong light. Crystals with plate or column shapes provide an excellent surface from which the light may reflect toward the viewer's eyes. Ice crystals in these forms can be found in ice clouds (*cirrus* or *alto* forms), ice fogs, snow virga (snow falling but not reaching the ground) falling from high-based clouds, blowing snow, and diamond dust. Since the light rays forming pillars are reflected rays, they take on the color of the incident light. For example, when the sun is higher in the sky, pillars are white or bright yellow. When it is near the horizon and its light color is dominantly orange, gold, or red, so is the pillar.

The formation of hexagonal-plate crystals is favored at air temperatures from 0 to –4°C (32 to 25°F) and from –10 to –20°C (14 to –4°F). Ice crystal plates resemble dinner plates with a hexagonal pattern in their long dimension and are thin relative to their width. When they are less than 15 to 20 micrometres across, crystal plates tumble randomly through the air as they fall. When the plates are larger, they fall so that their long dimension parallels the ground, floating down like flying saucers.

Ice columns resemble stubby pencils rather than the delicate branched snowflake shape our mind conjures at the mention of snow. Columns typically form in the temperature ranges –5 to –8°C (23 to 18°F) and below –25°C (–13°F). They are long in comparison to their hexagonal cross-section. Larger column crystals also fall with their long axis paralleling the ground. At times, the falling columns may rotate like slow miniature helicopter blades.

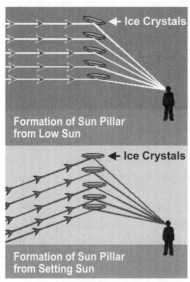

Geometry of formation of sun pillars from low (top panel) and setting (bottom panel) suns.

When the larger plates and columns assemble in a stable flying formation and strong light passes through the formation, the ice crystals act like small mirrors. Pillars appear when the light rays reflect at a grazing angle as they hit the crystals' upper or lower surfaces.

The breadth, form, and location of a light pillar depend on the type and orientation of the ice crystals, their height in the sky and distance from the observer, and the elevation of the light source. Plates roughly 1 millimetre (0.04 in.) across align their long axis nearly perfectly in

46

the horizontal plane as they fall and thus form narrow, bright pillars. If crystal alignment deviates slightly from perfect alignment, the resulting pillar broadens and may become detached from the light source. Ice plates generally produce sun pillars visible on the ground only when the sun is within six degrees of the horizon. Light reflected off ice columns can form pillars visible on the ground when the sun is higher, but rarely when it is more than twenty degrees above the horizon.

When the ice crystals, such as those within or falling from a cloud, are at a great distance from us, some slight misalignment of the crystals from the horizontal is required for pillars to appear. Otherwise, light rays reflecting from distant, perfectly horizontal plates would reflect away from our line of sight or simply pass through the crystal to our eye, not forming a visible pillar. When crystals are tilted slightly, the light can reflect off the lower outer crystal surface (or the inner upper surface) toward our eyes. When the sun is very low in the sky (one or two degrees above the horizon), however, reflections can occur off the lower surface of perfectly horizontal crystals. Since such pillars are often very narrow, their subtle structure can be lost amid the brilliance of a golden or crimson sunset.

For closer sources, such as street lighting, plate orientation is not as critical in pillar formation. When the crystals have a wide range of alignments, they form a much broader pillar, giving it a sheet-like appearance. Such pillars can vanish and reform with amazing quickness if gusty winds rapidly rearrange the crystal alignments, thus giving the impression of a shimmering aurora overhead. While sun pillars are rarely more than twenty degrees in height, pillars forming over artificial light sources are not restricted to this height angle. Such pillars may appear as tall as ninety degrees, depending on the reflecting crystals' height and observer's location.

When the ice crystals are very high in the sky—in clouds or snow virga—only reflections off the bottom of the crystals will form pillars visible to an observer on the ground. However, if viewed from a mountaintop or aircraft, pillars formed by reflections from the crystals' upper surfaces can be seen. From such vantage points, pillars extending below the light source are common. When ice crystals are present near the surface, such as during light falling snow, diamond dust, and blowing snow, reflections may be seen from the upper surface of some crystals as well as the lower surface of others. This produces pillars that extend both above and below the light source. Street lighting can also form double pillars.

Sun and moon pillars are not restricted to the cold season since high-altitude cirrus clouds, which contain ice crystals, are common in all months. At times, light pillars occur simultaneously with other

Ice Plates   Ice Columns

Plate ice crystals and ice columns.
(NATIONAL OCEANIC AND ATMOSPHERIC ADMINISTRATION/
U.S. DEPARTMENT OF COMMERCE, NOAA HISTORIC NWS
COLLECTION, PHOTOS BY WILSON BENTLEY)

optical phenomena generated by ice crystals, such as halos and sun dogs. Blowing snow, ice fog, and diamond dust can not only produce interesting pillars with sun or moonlight, but can also generate interesting effects around strong artificial light sources.

The most brilliant pillars I witnessed appeared one moonlit evening when I was cross-country skiing in a Guelph, Ontario, park. A blustery wind lofted sheets of snow off a small moraine and across the open field in front of me. Opposite the trail I was skiing, mercury vapor streetlights lined the boulevard. On this magical evening, each of those lamps wore a pillar above and below its bulb, an evenly spaced queue of vertical dirty-yellow feathers stretching before me.

Twelve

Icicles

Icicles are one of the visual symbols of winter, a reminder of the frigid bite of winter cold locking water into solid, stoic ice. In its basic form, an *icicle* is a tapered, hanging spike or cone of ice formed by the freezing of dripping or falling water. We picture icicles hanging off houses and other buildings, as water from roof-collected ice and snow melts, runs off, and is quickly recaptured by winter's frosty breath. Icicles can also form on trees, utility poles and fences, on rocks near waterfalls, or at ground water seepage points in rock cliffs. They result not only from dripping meltwater, but also rain, mist, and sea spray.

Icicle formation requires liquid water, subfreezing temperatures, and the pull of gravity to draw the shape into a candle-like taper. The first freezing of the down-flowing water becomes the icicle's root. An icicle grows as water continues to drip over the root, freezing in progressive layers. If multiple liquid water streams are separated, individual icicles form. When the liquid streams are close or heavy, the icicle spikes become broader-based ice appendages, at times forming a frozen cascade, such as those seen near waterfalls.

Icicles and sunlight can produce magical visages of glitter and gleam as light passes through or reflects off the ice surface in a variety of angles. There is a world of beauty in icicles if we study them up close. First, we are likely to notice the ribbed outer surface that is not smooth like a dinner candle, but molded in frozen undulations. When icicles grow, they do so downward and outward, simultaneously but at differing rates, thus forming a series of vertical ridges and horizontal ribs or rings on their outer surface. Close inspection also reveals bubbles frozen within the ice.

Surrounding the tip, we can see that it is not a solid mass but actually a delicate hollow in the ice filled with liquid water, dangling a pendent drop from the end. The liquid generally extends no more

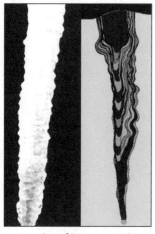

Layering of ice as an icicle grows. The right panel shows successive layers in the growth of a large icicle. (KEITH C. HEIDORN)

than a few centimetres up into the icicle interior. Occasionally, an air bubble enters this cup and drains it of liquid. During active growth, the water flowing down the exterior taper restores the contents to the inverted cup, often trapping air bubbles in the process. The greater the number of trapped air bubbles in an icicle, the more milky the ice appears as they diffuse and scatter light passing through the transparent ice.

An icicle may undergo a series of freeze-growth cycles over its lifetime, as long as there is liquid water available to add. Vertical ridges on the icicle form during renewed growth after a period of dormancy as meltwater streams down the exterior, initially laying a thin track of ice and adding to the rumpled appearance. Wind flowing past the icicle may shift the melt stream's track and initiate new channelling and ridging.

Even when active growth has ceased, an icicle continues to change shape and appearance, even at subfreezing temperatures. Some ice may sublime from solid to water vapor state, slowly altering surface features, and wind and precipitation can erode or break the icicle body. An icicle can grow to several metres in length over several days before melting, gravity, winds, or a child with a stick eventually bring an end to its existence.

Icicles forming on or near waterfalls, and other areas where liquid water seeps from soil or rock cuts, add beauty to natural winter scenes. In some cases, the spray from falling water drifts onto existing icicles and forms an extensive intertwining network of merged ice and hanging spikes, looking like an ice curtain pulled over the area. The largest icicles I have seen grew from the ever-flowing waters of Niagara Falls. These formed a *faux* falls, shadowing the animated Niagara cascade, even concealing it in places.

For home and building owners, those tapering spikes of ice hanging from roof edges are common winter problems. Large icicles can pull down gutters. During thaws or windstorms, falling icicles can be extremely dangerous to people and property below—it's not uncommon for police to close off busy downtown streets if a dangling icicle threatens from thirty stories up.

Thirteen

The Highs and Lows of Weather

On weather maps on television, in the newspaper, or online, we're told that the symbols "H" and "L" designate the current position of roaming regions of high and low pressure. However, *Highs* and *Lows* are actually very complex three-dimensional elements.[4]

Perhaps surprisingly, given the high attention paid to Highs and Lows, the absolute difference in surface air pressure between the two is usually much less than 10 percent of the total atmospheric surface pressure. Even the greatest extremes of sea-level pressure observed are less than 10 percent above or below the average sea-level value for Earth: 101.3 kilopascals (1,013 mb). However, when we ascend in the atmosphere, whether starting in the center of a High or a Low, or anywhere in between, the absolute air pressure decreases because less atmosphere remains above to press down on us. This is why we must discuss the pressure field of surface Highs and Lows as *sea-level pressure*. If surface readings were not corrected for their elevation above sea level, weather maps would be chaotic and offer little enlightenment as to the state of the weather. For example, Calgary would have a perpetual low-pressure region around it compared to Vancouver or Halifax, where absolute pressure would always be higher. Toronto and Winnipeg would sit on a shelf somewhere between. Generally, weather reports of pressure refer to sea-level pressure.

The High

Let's start with the high-pressure cell. Highs are characterized by sinking air at their center and mostly precipitation-free, clear skies.

---

4 I refer to Highs and Lows as sea-level features as plotted on a standard surface weather map. For the most part, I will consider them as quasi-two-dimensional elements. For convenience, I capitalize these terms when they refer to the complete entity.

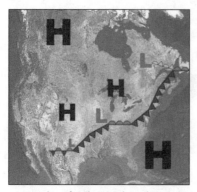

A cyclone family stretches along a frontal boundary across North America from the Canadian Maritimes to the American Southwest.

The highest pressure is found at its central core and pressure decreases outward from there. In the central core of the high-pressure region, the air is descending. When it reaches Earth's impenetrable surface, there is a temporary build-up of air mass before it can flow away from the High's core. This added mass exerts a higher force, or weight, on the surface below. Since *pressure* is defined as a force exerted over an area, the force of the additional mass results in a relatively high central-pressure zone.

As a result, a *pressure gradient* (the change of pressure with horizontal distance) pushes outward from the center toward the edges, causing the air to move. Moving air, of course, is wind. If Earth did not rotate, this wind would blow directly outward from higher to lower pressure. But the planet does spin, and this affects the air's long-term trajectory, an influence called the *Coriolis force* or *Coriolis effect*, which is strongest in the polar regions and weakest in equatorial latitudes. In the northern hemisphere, the Coriolis effect causes the wind to turn to its right as it blows over Earth's surface. Thus, in the ideal scenario, winds would circle the high-pressure cell in a clockwise manner (a pattern termed *anticyclonic* because it is opposite in direction to flow around a cyclone/low-pressure cell) as the Coriolis force balances the pressure gradient force, a flow known as *geostrophic flow*.

Over the oceans and flat land areas (mountains greatly complicate weather patterns), this geostrophic balance holds quite well above the surface layer (the lowest kilometre of atmosphere). Within the surface layer, however, friction also plays a role by altering the balance between pressure gradient and Coriolis forces to establish a new balance known as *gradient flow*. Gradient flow winds cross the pressure gradient such that the air flows slightly outward from the high-pressure cell core.

Following their initial formation, most Highs take a generally elliptical shape and are often large and sprawling. But as they interact with other air masses and topography and are distorted by forces of the upper atmosphere, high-pressure cells often become long and narrow. When plotted on a surface weather map, these elongated pressure patterns resemble mountain ridges on terrain maps. Meteorologists therefore refer to them as *high-pressure ridges* or simply *ridges*.

Since the air in a High sinks through a deep layer of atmosphere, the formation of clouds and precipitation is greatly inhibited. Generally if clouds do form, they are spawned by strong solar heating of the surface and are small and of limited vertical extent—usually fair weather cumulus (*cumulus humilis*). If the air of the High is humid and very warm, such as found in tropical air masses, and solar heating is strong, isolated showers may form within the High, usually along the cell's outer edges. Otherwise, unless topography throws a wild card into play, high-pressure regions are generally precipitation free.

In some situations, however, a layer of warmer and moister air may be pushed far into a High by high-altitude winds. In these situations, high *stratiform* (layered, sheet-like) clouds may appear distant from any obvious surface weather front. Meteorologists have begun calling such situations "dirty Highs" because they appear as grayish, dirty regions on satellite imagery.

The Low

A low-pressure cell is characterized by converging airflow at Earth's surface, ascending air currents at its core, where the converging currents meet, and regions of divergent airflow above where air currents flow away from the rising core. When the combination of these processes removes air aloft faster than it can be replaced by the converging low-level inflow, the temporary mass deficit in the air column lowers its weight, reducing the surface pressure to lower than that of the surrounding atmosphere.

To replace the ascending air at the Low's core, surface air must flow toward the cell's center. But as with the circulation out of a high-pressure cell, the Coriolis effect prevents a direct flow route. Thus, the geostrophic balance around a low-pressure cell produces a wind field that flows around the Low in a counterclockwise direction in the northern hemisphere, called a *cyclonic flow*. When surface friction affects the flow, the resulting gradient wind has an inward component toward lowest pressure. Ascending currents within a Low push surface air toward higher altitudes. Air moving upward cools by expansion, usually reaching its condensation level along the way. Clouds form at the condensation level and extend upward in the atmosphere. If conditions are right, precipitation develops as more and more water vapor is transformed to liquid water or ice crystals that eventually fall earthward. Thus, the presence or approach of a low-pressure cell generally signals cloudy and wet weather ahead.

When a quasi-circular region of low pressure elongates to a long and narrow band, it is referred to as a *trough of low pressure* or just a *trough* (as opposed to a high-pressure ridge). Low-pressure cells that travel long distances across the earth are called *cyclones*, or *storm systems*, due to their attendant stormy weather.

Preferred Locations for Low Formation

The formation of a low-pressure cell, called *cyclogenesis*, can be complicated. Some Lows form in specific geographical locations. Others form through the clash of contrasting polar and tropical air masses and the influence of high-speed air currents aloft, which give a twist to the mobile *polar front* zone between two air masses.

The leeward sides of mountains are common low-pressure spawning grounds. Over North America, cold-season storms such as the Colorado Lows and Alberta Clippers emerge from the lee (east) of the Rocky Mountain chain. During the summer above desert regions, extreme solar heating may form large, quasi-stationary *thermal lows*.

Large bodies of water, such as the Great Lakes, can be breeding grounds for low-pressure systems or assist in the development of weak lows moving over them. Cyclogenesis over water occurs when the water mass is significantly warmer than the air passing over it. When heat from the warm water is imparted into the system, updrafts are formed and the development process begins. Over the Great Lakes, particularly Superior and Huron, autumn is the peak season for Low formation.

There are also several cyclone spawning regions around the globe. One, the Aleutian Low off Alaska, spawns the storms that worry North America's northwestern coast each winter. Its counterpart off Iceland sends storms into Europe. Tropical regions too have preferred locations that produce tropical storms, hurricanes, and typhoons during specific periods of the year. In the Atlantic Ocean, areas west of the Azores and in the Gulf of Mexico are preferential locations for tropical storm and hurricane development from June through November.

Frontal Low Formation

In the middle latitudes, many low-pressure cells and cyclonic weather systems develop in the zone between two contrasting air masses (Highs). The zone between air masses of polar origin and those of tropical birth is termed the *Polar Front*—the explanation for this storm development process is known as the Polar Front Theory of Cyclonic Development. Developed by Norwegian meteorologists about a century ago, this theory was a significant step in understanding weather and weather forecasting. It also gave us the now-common term *front* for the boundary between contrasting air masses.

In the polar frontal zone between the air masses, the complex process of frontal low formation begins. Simplistically, it usually starts when some combination of surface and upper-level factors causes the frontal boundary between the air masses to kink. This kink, known as a *frontal wave*, pushes cold air under warm air, forming a *cold front*, and warm air over cold, forming a *warm front*. This causes air to converge and then ascend around the kink. If upper-air conditions are right, the ascending air produces a region of lowering pressure, and a spinning cyclonic system with characteristic frontal boundaries begins to develop. These Lows are called *frontal lows*, and the inverted V-shaped frontal region around the Low is termed the *frontal wedge*.

Eventually, the contrasting air masses around the low-pressure cell mix well enough to weaken their contrast, ending the cyclone's driving force. Its storminess weakens and its pressure rises until the cyclone eventually disappears from the weather map.

One cyclone's death, however, may give birth to another farther downstream. As an old storm dissipates, a new cell may spin up along the western tail of the frontal zone. A chain of Lows in various stages of development along the polar front is called a *cyclone family*. During the winter, a well-developed cyclone family can often be seen sprawling along the polar front stretching from Newfoundland to west Texas.

A large low-pressure system centered over Hudson Bay covers parts of Manitoba (left), Ontario (bottom center), and Quebec (right). (JACQUES DESCLOITRES, MODIS LAND RAPID RESPONSE TEAM, NASA)

Fourteen

Air Masses: A Fine Vintage

Years ago, I wrote a poem called "A Fine Vintage" in which I used the metaphor of wine tasting and variations in vintage to describe different air masses. I described how our senses of smell and sight, and at times taste, can distinguish the type, origin, and travels of the air masses surrounding us at any given moment. The contrasts between air masses—say, stifling hot and humid maritime tropical air and chillingly cold, dry polar continental air—provide one of the clues weather forecasters use when analyzing weather maps to produce forecasts.

Air masses are distinct in their temperature and moisture content due to their place of birth. A number of such breeding grounds ring the globe, tucked away from strong global wind belts so that the nascent air mass can mature. Eventually, all air masses are pushed from their nest by global wind currents forcing them to mingle with other traveling air masses. Cold air masses generally seek warmer latitudes, while warm ones prefer northward journeys.

Good weather forecasters can build a forecast on a solid knowledge of the properties of the various air masses around their forecast zone in much the same way a wine taster can "forecast" the popularity of a good wine vintage. Since the air masses of concern often are hundreds or thousands of kilometres away, weather forecasters must rely on a variety of measurements (such as temperature and humidity) to tell them the "taste," "bouquet," and "color" of the various air masses.

Air Masses Defined

An *air mass* is a large dome of air that has similar horizontal temperature and moisture characteristics. At any given time, an estimated fifty distinct air masses are scattered across the planet. Some are newly born entities that strongly reflect their birthing ground. Others are old and travel-scarred, with only the smallest commonality with their place of origin remaining.

Distinctly different air masses are separated by narrow transition zones that weather analysts distinguish by drawing weather fronts—warm, cold, stationary, and occluded fronts—between them. Sometimes these fronts indicate such subtle transitions, perhaps only a shift in wind direction, that they are hardly noticed as they pass. Others are vigorous zones where conflicts between warm and cold air masses produce very heavy weather—severe thunderstorms or heavy snow. Often, the conflict between air masses is so intense that great cyclonic weather systems develop along the frontal boundary to lash the surface beneath with high winds, rapidly changing temperatures, and precipitation of all varieties. It is no wonder the Norwegian meteorologists who developed the concept of air mass analysis and their frontal boundaries used terms from warfare to describe the situation.

The differences among air masses were likely first recognized when humans realized that major changes in weather had distinguishable, repeatable patterns. For example, cold, dry conditions come from the north; hot and dry, or hot and humid weather from the south. But, until the seventeenth century, early *meteorologists*, like our wine tasters, had to use their senses to distinguish the difference between them. In that century, the invention of instruments, such as the thermometer, barometer, and hygrometer, allowed observers to measure objective properties of the air. With the advent of regular weather observations in the latter half of the nineteenth century across large regions of the continents, meteorologists began to see repeatable patterns that indicated large bodies of air could be distinguished by their temperature and humidity levels.

The first formal theory of the impact of air mass differences came out of the famed Bergen (Norway) School of Meteorology during the early decades of the 1900s. Led by Vilhelm Bjerknes, the research group, which included his son Jacob, Halvor Solberg, and Tor Bergeron, laid the foundation for modern weather analysis and forecasting. The group developed the concepts of frontal analysis, wave cyclone formation, and air mass analysis, to name a few of their achievements.

One significant finding was contained in Bergeron's 1928 doctoral dissertation. There, he confirmed that certain characteristics of air masses did not substantially change or age for long periods of time as the air mass flowed over oceans and continents. Therefore, Bergeron concluded, knowledge of these characteristics was fundamentally important to improving weather forecasts.

Bergeron saw air masses as being of four types: Equatorial,

Tropical, Polar, and Arctic (or Antarctic). From this, he developed an elaborate classification scheme that included distinguishing properties of temperature, humidity, and aerosol content (as measured by visibility). With slight modifications, his classification system remains a viable concept today in weather analysis and forecasting.

Air Mass Classification

Most air masses cover thousands of square kilometres of surface and extend several kilometres vertically. Each one bears the mark of the region in which it was formed, to some degree or other. Some of the fifty are young and fresh. Others are old and greatly transformed. Some are moving across the planet at speeds covering several hundred kilometres each day, others are nearly stationary.

Air masses acquire their characteristic temperature and moisture (or absolute humidity) signature from the *source regions* over which they are born. The ideal source region is one with light winds, particularly in the upper atmosphere, so that the air mass remains in place long enough to acquire the temperature and moisture properties of the underlying surface throughout its mass. Therefore, middle-latitude regions, where weather systems move quickly across the surface, driven by fast-moving upper-level air currents such as the *jet stream*, are not good air mass breeding grounds.

There are several source regions of extensive, semi-permanent high pressure around the globe, in particular, within two latitude belts in each hemisphere: one in the polar regions, the other in the subtropics. At varied intervals, portions of these semi-permanent high-pressure cells break away to form vagabond air masses that hitch a ride on strong upper air currents and travel the globe.

High polar latitudes and the subtropics around 30° latitude are both good source regions, whose relative strength waxes and wanes with the solar seasons. Open ocean expanses, large deserts, and extensive continental plains at high or low latitudes make the ideal birthing grounds within these belts. Mountainous areas are too variable in their properties, and midlatitude continental plains are not conducive to air masses staying in place for long because of the strength of the prevailing westerly global winds at these latitudes. Under certain conditions, midlatitude oceans can be source regions because their surfaces have very uniform characteristics.

From the characteristic properties picked up in their breeding ground, air masses are designated as hot or cold, wet or dry. The terms are to some degree relative. A cold air mass in summer may be as warm as a warm air mass in winter. Each air mass has a characteristic temperature and moisture content and thus we can distinguish four combinations: hot and dry; hot and wet; cold and wet; and cold and dry. Bergeron included two additional temperature categories by defining "very hot" and "frigid" air masses for those forming over the equator and polar regions, respectively.

Wet air masses are air masses formed over the oceans, and *dry* air masses, those formed over the continents. *Equatorial* air masses are all considered to be wet because much of the land area under the equatorial zone is covered in tropical rainforests that can add as much moisture to the air as the equatorial oceans. All *Arctic* (or *Antarctic* in the southern hemisphere) air masses are considered dry because there is little evaporation into them from the frigid polar oceans and their temperatures are so low that even at saturation, the absolute humidity is very low.

The first dimension of the Bergeron classification system is the latitude zone of air source region, which governs the air mass's temperature characteristics. There are four such zones in the system: Equatorial (E), Tropical (T), Polar (P), and Arctic or Antarctic (A or AA). (The letter in parentheses is the corresponding label used on weather maps.)

Next are the two underlying surface characteristics of the source region that affect the resulting air masses: Maritime or Oceanic Surfaces (m), which create relatively humid air masses, and Continental or Land Surfaces (c), which create relatively dry air masses.

The combination of the above yields eight air mass types, but because Arctic (Antarctic) and Equatorial air masses have only one moisture character, there are six basic air mass types, given below:

Air Mass Types and their Characteristics

Air Mass Type	Temperature Characteristic	Moisture Characteristic
Arctic or Antarctic (A or AA)	*Extremely cold*, formed over poles	*Very dry* due to extreme cold
Polar Continental (cP)	*Very cold*, having developed over subpolar regions	*Very dry*, due to the cold and having developed over land
Polar Maritime (mP)	*Very cool* because of the high latitude but not cold, due to the sea's moderating influence and the warm ocean currents at these latitudes	*Moderately moist* because of the cool temperature, but not as dry as polar continental air because of evaporation from the water surface
Tropical Continental (cT)	*Very warm* because of the lower subtropical latitude of formation	*Dry* because it formed over land
Tropical Maritime (mT)	*Very warm* because of the subtropical latitudes at which it forms	*Very humid* because of the warm tropical waters below
Equatorial (E)	*Hot*	*Extremely humid* whether formed over land or water

On some weather maps, the lowercase letters "k" or "w" may be attached to the two-letter abbreviation describing an air mass. The "k" indicates that the air moving across a region is colder than the land surface temperature, while the "w" indicates the air is warmer than the land surface temperature. Thus, cold continental polar air flowing over warmer land surfaces would be designated "cPk."

Air Masses Affecting Canada and the United States

Five of the six basic air mass types affect weather in Canada and the continental United States. They can bring anything from scorching heat to bone-chilling cold depending on the type of air mass and time of year. The most violent weather usually occurs during spring, when cold, dry continental polar air clashes with hot, humid maritime tropical air.

Arctic (A): Arctic air masses usually originate north of the Arctic Circle, where winter days of twenty-four-hour darkness allow the air to chill to extremely low temperatures. Such air masses break southward across Canada and the United States during winter, but are very rarely seen at lower latitudes during the summer because the twenty-four-hour sun warms the Arctic region considerably, and the polar front and accompanying jet stream generally remain at higher latitudes.

Continental Polar (cP): Cold and dry, continental polar masses are not as cold as arctic air masses. These usually form farther south, in the subpolar Canadian north and Alaska, and often dominate the weather picture across the continent during winter. Continental polar masses do form during the summer, but mostly influence only Canada and the northern United States. These air masses are usually responsible for bringing clear, pleasant weather in summer.

Maritime Polar (mP): Cool, moist conditions characterize maritime polar air masses. They usually bring cloudy, damp weather. Maritime polar air masses form over the northern Pacific and the northern Atlantic oceans. These generally influence the Pacific Northwest and the Northeast, respectively. Maritime polar air masses can form any time of the year and are usually not as cold as continental polar air masses in winter because of the moderating influence of the sea surface beneath them.

Maritime Tropical (mT): Warm temperatures with copious moisture typify maritime tropical air masses. They are most common across the eastern United States and southeastern Canada, originating over the warm waters of the southern Atlantic Ocean, Caribbean Sea, and the Gulf of Mexico. These air masses can form year round, but they are most prevalent during summer. Maritime

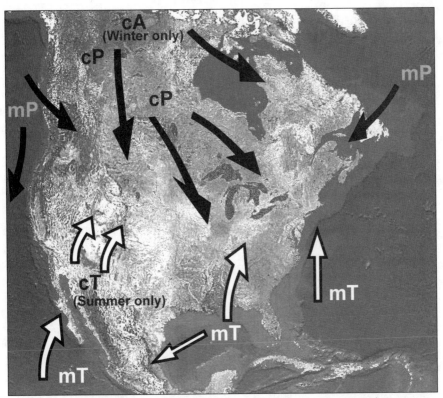

Source regions for common North American air masses. (mT—maritime tropical air mass; cP—continental polar air mass; cA—continental arctic air mass; mP—maritime polar air mass; cT—continental tropical air mass)

tropical air masses are responsible for the hot, humid days of summer across much of the eastern half of the continent. Such air masses are often called the *Bermuda High* because of their birthplace within the subtropical zone around and east of Bermuda.

Continental Tropical (cT): Hot and very dry, continental tropical air masses usually form over the Desert Southwest and northern Mexico during summer, often keeping the region scorching above 38°C (100°F) during summer. They can bring record heat to the U.S. plains and the Mississippi Valley during summer, but they usually do not make it to the eastern and southeast United States or into Canada as cT air masses. As they move east, moisture evaporates into the air, transforming the air mass to become more like a maritime tropical air mass. Continental tropical air masses very rarely form during winter.

Equatorial (E) air masses rarely visit the contiguous United States and almost never reach Canada, but they are an important weather factor for the southern nations of North America, including Mexico, Central America, and the Caribbean.

Take It Outside

The next time you view a weather map, look at the areas of high pressure and see how each presents its unique face. Think of them as different wines: warm air masses as red wine, cold air masses as a crisp white—with varying degrees of sweetness (humidity). Or just think of them as having a color: red for hot and humid, orange for hot and dry, blue for cold and dry, and green for cold and moist. Then go out and sense the look, feel, and even smell of the air mass over you. After a while, you will develop a new appreciation for air masses, which have more character than the weather map's rather simplistic "H" suggests.

SPRING

DUST DEVILS • SNOW PILLARS • BLIZZARDS • CLIPPERS • SNOWFLAKES • CREPUSCULAR RAYS • MOONBOWS • HALOS • SUNDOGS • RAINBOWS • LIGHTNING • THUNDERSTORMS •

Fifteen

20 March: The Vernal Equinox

The *vernal equinox*, also called the spring equinox, is to many in the northern hemisphere the first day of spring. For those following the solar calendar, however, it is actually Midspring Day. This is one of the two days of the year when the sun is overhead at the equator, and all inhabitants of Planet Earth share a day with equal length, at least in theory—all should enjoy twelve hours of day and twelve hours of night, except the atmosphere plays a trick on us, as we shall see shortly.

Spring provides the transition from winter to summer; however, its astronomical start date is by no means reflected in the appropriate changes in weather and climate. The operative word is *transition*, and the change is rarely a smooth one, though it can be a rather steady one. In reality, of course, spring's beginning varies from place to place, and year to year, and even from person to person, depending on what standard is used to mark the beginning. For some it is marked by the sight of the first robin, or the first cherry blossom (which pops out on southern Vancouver Island in early to mid-February), or the first picnic, or first time out on the lake.

However you choose to define spring, most cultures celebrate winter's end as a time for rebirth and renewal. The ancient Greeks celebrated spring with the Festival of Flowers in honor of Dionysus, the youthful Greek god of wine and resurrection. The Romans participated in a number of spring festivals, including Bacchanalia. Christianity celebrates the season by marking the resurrection and rebirth of Jesus after a symbolic period of denial, Lent, which is similar to the resurrection and rebirth of spring following the deadness of winter.

The season's English name has deep roots. "Springtime" first appeared in text around 1398. The name alludes to the season as a time when the world springs back into action after winter's rest and new life springs forth from earth, water, and womb. It is also the time when the sun springs back over the equator for residents of the northern hemisphere, in contrast to its fall below in the autumn. Sometime later, "spring of the leaf" was used to refer to the season, and eventually the name was shortened to spring.

We have long been taught that on the equinox the lengths of day and night are equal, twelve hours each. For most practical observations, this is true and would also be absolutely true if the planet had no atmosphere. But the sunrise and sunset tables give a twelve-hour separation several days before the equinox and report a longer day than night by several minutes on the equinox itself. A grade school teacher asked me about this when she was having her class watch the newspaper sunrise/sunset times as the autumnal equinox approached. She was taken aback that they did not agree with her traditional teachings and asked me why.

The spring equinox occurs on the day when the sun is directly overhead at the equator. From an astronomical/geophysical viewpoint, the day and night lengths *should* equal twelve hours each when this occurs because the solar rays beam tangentially on both poles. But the geometry works perfectly only if a planet has no atmosphere or a very, very thin one.

The lengths are not the same because Earth's atmosphere is thick and dense enough to refract (bend) the rays of light coming from the sun in such a way that the sun appears on the horizon several minutes before it actually arrives there, in essence, producing a *superior mirage* (which I discuss in detail later in this book on page 94). A similar situation occurs at sunset. When the sun appears to be on the horizon, it is actually already below it, having set minutes earlier. However, the official definition for sunrise and sunset used by the various astronomical observatories, such as Greenwich Observatory, is the moment when the upper edge of the sun's disk is on the horizon, considered unobstructed relative to the location of interest. For the preparation of sunrise-sunset tables, the atmospheric conditions causing the refraction are considered average, so in fact, the sunset and rise times on the equinox may vary slightly with the overall weather conditions. These extra minutes at sunrise and sunset account for the slightly longer day length on the day of the equinox. The day when the sunrise and sunset times are exactly twelve hours apart occurs about three days earlier in the spring, when the true solar position would give a day length of less than twelve hours.

For observers within a couple of degrees of the equator, the period from sunrise to sunset is always several minutes longer than the night. At higher latitudes, the day of the equinox is about seven minutes longer than the night, increasing to ten minutes or more at

50° latitude. The specific dates of equal day and night are also different for different latitudes. In the northern hemisphere, at 5° latitude, the dates of equal day and night occur about 25 February for the spring equinox and at 40° latitude on about 17 March.

Sixteen

The March of Spring

Spring comes early to those of us residing in and around Victoria, British Columbia. We annually revel in its early appearance and gloat over our reputation as the "Canadian tropics" by taking an annual blossom census in late February while our continental Canadian cousins still freeze and shovel snow. The regular early onset of spring can be credited, in part, to our mild winters, a gift from the Pacific Ocean's relatively warm waters lapping at our shoreline.

I often first note the emerging cherry blossoms in my neighborhood during February's first week, but in the city's southern neighborhoods—about 5 kilometres (3 mi.) away—many cherry trees have already reached full bloom. By Valentine's Day, the street outside my window is frequently lined with white-blossomed cherry trees. During my decade at this location—5 kilometres (3 mi.) from the shoreline and 60 metres (197 ft.) above sea level—I have noticed that boulevards lined with cherry trees burst into color like a slow flood tide moving from the oceanside streets to my part of town, creeping a few blocks closer every day. That is spring's migration on the microscale here in Greater Victoria, but it is not the only one taking place. Each year, spring begins migrating north along the eastern/Atlantic regions of North America from the ever-warm reaches of south Florida toward the Canadian Maritime provinces, eastern Quebec, New England, and beyond. Similar marches roll up the mid-continent.

While people appear to want spring to start "on the dot" with the equinox, nature does not work that way. Not all seasonal rhythms follow precise clocks, even if the planet's orbital position does. Most often, they build and recede across continental areas like the flow and ebb of ocean tides. Spring illustrates such natural, seasonal tides perfectly because it has dramatic, visible indicators in its

Spring cherry blossoms come early in Victoria, British Columbia (February 2003). (KEITH C. HEIDORN)

changing flora and fauna. Though spring's northward migration does not advance with the speed of a warm front, its climatological advance is very apparent to nature watchers. New plant life emerges or blossoms, and animals migrate, hatch, or awake from winter sleep.

Studies have established that spring migrates northward along the coast of eastern North America at about 27 kilometres (17 mi.) per day at sea level. Thus, if red maples bud in Washington, D.C., on 20 March, trees in Baltimore should bud two days later, with New York City's red maples budding two weeks hence. This rate applies only to spring's sea-level advance northward. Spring must also climb the terrain, doing so at a rate of around 30 metres (98 ft.) of altitude per day. Therefore, once spring arrives in the valley, it will finally reach the top of a 300-metre (984-ft.) ridge ten days later.

The Science of Phenology

We recognized this northward migrating tide even before regular and widespread measurements of weather conditions began in the nineteenth century. Much of that knowledge came through long-term observations of surrogate climate indicators: plants and animals. Such study constitutes a branch of science known as *phenology.*

Phenology is the science dealing with the relations between climate and periodic biological phenomena. In other words, it is the study of cyclic events of nature—usually the life cycles of plants and animals—in response to seasonal changes in their environmental conditions. Such changes may be variations in the duration of sunlight or changes in temperature. Phenology is a prime example of what important scientific information can be produced by simple observation and record keeping—though the first "phenologists" likely had no interest in science, they did it to assure a full belly. All it really takes to be a phenologist are good senses for observation, a diary or journal to record observations, and time—several years.

Any of the many cycles in the living world can be watched and recorded in a date-keeping journal to start a phenological study. Among the most popular areas of study are:

- plant budding and floral blooms in the spring, summer, and autumn
- spring and autumn migration patterns of certain bird and mammal species
- denning and emergence dates of hibernating animals
- appearance of fireflies, mosquitoes, and other insects, both pests and beneficials
- fruiting and harvest dates of various cultivated plant species

Over a number of years, these records provide clues to the timing of biological cycles, such as those applicable to hunting, gathering, and agriculture. Today, we base many similar decisions, such as the expected crop harvest date, on observed weather data and computer models. Our ancestors watched the natural world around them for such decision-making signs.

Phenology is an old science. There are written records from as early as 974 BC, but the first phenological studies likely came early in human evolution. To a large extent, early human survival was likely linked with the ability to understand and act upon the clues given by seasonal changes. The ability to make the right decisions based upon phenological observations might have been the difference between life and death. Some Native Americans, for example, observed that when oak leaves reached a particular size, it was time to plant corn seeds. If corn was planted at this time, the soil was warm enough to prevent corn seeds from rotting, and a sufficient stretch of warm weather remained ahead for corn plants to fully mature before cold weather again set in.

Carolus Linnaeus, the founder of our plant/animal classification and naming system, initiated the first systematic phenological studies in 1750, when he selected eighteen sites in his native Sweden and made yearly observations of the dates of leaf opening, flowering time, fruiting, and leaf fall, together with records of weather conditions.

While anything in the biosphere that has recurring rhythms is a prime candidate for phenological study, other natural events can also be watched for clues of regularity. In weather, the dates of last snow or last frost of spring, first snow or first frost of autumn, or the dates a local lake or pond freezes in the fall or opens in the spring are candidates for study. The ancient Egyptians, for example, found that the flooding of the Nile came annually around the time when Sirius, the Dog Star, rose just before the sun.

Earlier, I gave values for the march of spring northward along the North American East Coast and upwards with elevation. Those figures were first derived by America's leading forest entomologist during the first part of the twentieth century, Andrew Delmar Hopkins. When Hopkins looked at the phenological records for Virginia and West Virginia, he found a relationship between the coming of spring and a site's elevation, latitude, and longitude. Applying this farther afield, he discovered that in North America, the farther west, or north, or higher in elevation from the Atlantic coast a site was, the later spring arrived. The relationship, today known as *Hopkins' Law of Bioclimatics*, states that spring advances one day for every fifteen minutes of latitude northward, one and a quarter days for each degree of longitude westward, and one day for every 30 metres (98 ft.) higher in elevation. Thus, for a site located fifteen minutes of latitude north, one degree west and 60 metres (197 ft.) higher than another chosen site, spring should arrive 4.25 days later.

Phenology and Climate Change

Phenology can also provide important information about climate change, at least giving a broad marker independent of temperature measurements to judge whether a change is or is not occurring. By comparing data from past studies with those of today, bioclimatologists are seeing changes in the timing of biological events, indicative of a warming climate in many parts of the world. For example, spring flowering in Alberta is now as much as ten days earlier than it was forty-five years ago. British researchers report that the first spring flowering of 385 British plant species has advanced between four and a half days and fifty-five days in the last decade compared to the flowering date of those species between 1954 and 1990. The first flowering date in the 1990s had advanced by an average of fifteen days for 16 percent of the studied species. The greatest advance in the flowering date came with the white dead nettle, which advanced fifty-five days.

The lengthening growing season can have a different effect on those plants that depend on the number of cool days (days with mean temperature below a specific threshold) for their dormancy. Many of these plants now bloom later in the season because they need extra cool days to initiate growth. As a check on the hypothesis of a warming climate, plants that base their flowering on the length of daily light and dark cycles, rather than on temperature levels, have shown no significant changes in their blossom dates. While such studies cannot distinguish between climate warming caused by natural global climatic cycles or by human impacts on global climate, they do indicate that we are currently in a warming period.

Comes Spring

The arrival of spring, as indicated by plants and animals, or even meteorological measurements, does not mean that wintry weather will not return. Very often there are short-term setbacks and occasionally extreme relapses to winter weather. And it is rare that any indicator suddenly appears with the first burst of spring weather. What the bursting to life of plants and animals indicates is that a lengthy or consistent stretch of favorable, spring-like weather has occurred.

As spring advances across the continent, you can observe the march of nature in your own area. You might even be interested in beginning phenology as a pastime this year (and not only for spring events). Just pick some event, such as the first leaves, or buds, or robin's song, and record the date observed. The next year, watch again for that event and see if it is earlier or later. Or find an event occurring south of or at a lower elevation than your location, and see how long it takes for the event to reach you. Then you will have seen spring's northward (or upward) migration.

Making Clouds and Rain

With the coming of spring, warmer weather entices us outdoors more often and encourages us to take the time and watch the world go by. Perhaps "spring fever" strikes, and we are content to lie on a sunny hillside and watch the build-up of cumulus clouds over the landscape. We can look forward to April showers, rumbling thunderstorms, and cooling summer rains. But what makes clouds and rain?

Building Clouds

I'll start with clouds since they are necessary precursors for rainfall. Here is the basic recipe. First, we need two ingredients: water and dust.

On Planet Earth, naturally occurring clouds are composed primarily of water in its liquid or solid state. (On other planets, clouds may form from other compounds such as the sulfuric acid clouds found on Venus.) We begin the recipe by collecting a sufficient quantity of water in the vapor state. The water vapor content of the atmosphere varies from near 0 to about 4 percent, depending on the moisture on the surface beneath and the air temperature. To form clouds, the more vapor, the better.

Next, we need some dust—not a large amount, nor large particles, and not all types of dust will do. Without "dirty air," there would likely be no clouds at all, or only high-altitude ice clouds, where temperatures can sink extremely low. Even the "cleanest" air found on Earth contains about one thousand dust particles per cubic metre. Dust is needed for condensation nuclei, upon which water vapor may condense or deposit as a liquid or solid, respectively. Certain types and shapes of dust and salt particles, such as clays and sea salts, make the best condensation nuclei.

With proper quantities of water vapor and dust in an air parcel, the next step is for the air parcel mass to be cooled to a temperature at which cloud droplets or ice crystals can form. Voilà, we have clouds.

The simple recipe for making clouds and precipitation is a lot like cooking chicken: you take a chicken and some spices, apply heat, and after a time you have cooked chicken. But just as there are many ways to cook chicken, there are many different ways to form clouds.

Professor John Day, known as the Cloud Man, has expanded on the details of this simple recipe, calling it the *Precipitation Ladder*. As with our simple recipe, Day begins with the basic ingredients of water vapor and dirty air. In Rungs 3 through 8 of the "ladder" (see below), the ingredients move through several processes to form a cloud. Moving farther up the remaining three rungs leads to precipitation.

The Precipitation Ladder

11. Precipitation
10. Droplet Growth
9. Buoyancy/Cloudiness
8. Condensation
7. Saturation
6. Humidification
5. Cooling
4. Expansion
3. Ascent
2. Dirty Air
1. Water Vapor

Ascent, Expansion, and Cooling

Ascent and *expansion* are two of the main processes that result in the cooling of an air parcel in which clouds will form. We mostly think of moving air as wind flowing horizontally across the surface, but air moving vertically is extremely important in weather processes, particularly with respect to clouds and precipitation. Ascending air currents take us up the Precipitation Ladder. Where descending currents are present, we come down the ladder with processes reversing until we are finally left with water vapor and dust floating within an air mass.

There are four main processes occurring at or near Earth's surface that give can rise to ascending air: *convergence, convection, frontal lifting,* and *physical lifting.*

Convergence occurs when several surface air currents in the horizontal flow move toward one another to meet in a common space. When they converge, there is only one way to go: up. A surface low-pressure cell is an example of an area of convergence, and air at its center must rise as a result.

Convection occurs when air is heated from below by sunlight or by contact with a warmer land or water surface until it becomes less dense than the air above it. The heated air parcel will rise until it has again cooled to the temperature of the surrounding air.

Frontal lifting takes place when a warmer air mass meets a colder one. Since warm air is less dense than cold, a warm air mass approaching a cold one will ascend over the cold air. This forms a warm front. When a cold air mass approaches a warm one, it wedges under the warmer air, lifting it above the ground. This forms a cold front. In either case, there is ascending air at the frontal boundary.

Physical lifting, also known as *orographic lifting*, is found where horizontal winds are forced to rise in order to cross topographical barriers such as hills and mountains.

Whatever the process causing an air parcel to ascend, the result is that the rising air parcel must change its pressure to be in equilibrium with the surrounding air. Since atmospheric pressure decreases with increasing altitude, so too must the pressure of the ascending air parcel. As air ascends, it expands, and as it expands, it cools. The higher the parcel rises, the cooler it becomes.

Humidification, Saturation, and Condensation

Before an air parcel is ready to form a cloud, it must continue to cool until condensation is reached. The next several rungs of the Precipitation Ladder describe the processes through to the condensation of liquid water. As the air cools, its relative humidity increases—a process Day terms *humidification* (Rung 6). Although nothing has happened to change the air's water vapor content, the air parcel's saturation threshold decreases as the air cools. By decreasing the saturation threshold, the relative humidity increases. Cooling is the most important method for increasing an air parcel's relative humidity, but it is not the only one. Another is to add more water vapor through evaporation or mixing with a more humid air mass.

Humidification may eventually bring the air within the parcel to saturation. At saturation, the relative humidity is 100 percent. Before a cloud will form, a little more humidification is usually required, bringing the relative humidity slightly above 100 percent, a state known as *supersaturation*. When air becomes supersaturated, its water vapor looks for ways to condense out. If the quantity and composition of the parcel's dust content is ideal, condensation may begin at a relative humidity below 100 percent. If the air is very clean, it may take high levels of supersaturation to produce cloud droplets. But typically condensation begins at relative humidity a few tenths of a percent above saturation.

Condensation of water onto condensation nuclei (or the deposition of water vapor as ice on freezing nuclei) begins at a particular altitude known as the *cloud base* (or *lifting condensation level*). Water molecules attach to the particles and form cloud droplets, which have a radius of about 10 micrometres or less. The droplet volume is generally a million times greater than the typical condensation nuclei.

Clouds are composed of large numbers of cloud droplets, or ice crystals, or both. Due to their small size and relatively high air resistance, cloud droplets and ice crystals can stay suspended in the air for a long time, particularly if they remain in ascending air currents. The average

The sky shows a large vertical growth of cumulonimbus clouds with a dense lower level cloud deck.
(KEITH C. HEIDORN)

cloud droplet has a terminal fall velocity of 1.3 centimetres per second (0.04 ft./s) in still air. To put this into perspective, the average cloud droplet falling from a typical low cloud base of 500 metres (1,640 ft.) would take more than ten hours to reach the ground! The fall is so slow, it gives the impression that most clouds are floating in air. They usually evaporate in a much shorter time than it would take to fall down.

While clouds in their varied forms and appearances are a source of much interest, I will leave them for now and continue up the Precipitation Ladder toward the top rung: precipitation.

Forming Precipitation

We know that not all clouds produce precipitation that strikes the ground. Some may produce rain or snow that evaporates before reaching the ground, and most clouds produce no precipitation at all. When rain falls, we know from measurements that the drops are larger than 20 micrometres in diameter. A small drop of 1.2 millimetres (0.05 in.) diameter contains a million typical cloud droplets (0.012 mm [0.005 in.]). So if we are to wring some precipitation from a cloud, we need additional processes within the cloud to build raindrops from cloud droplets.

Buoyancy or Cloudiness

The next rung of the Precipitation Ladder is *buoyancy* or *cloudiness*, which signifies that the cloud water content must be increased before precipitation can be expected. This step requires a continuation of the lifting process. It is assisted by the property of water that releases heat when changing from vapor to liquid and solid states, the latent heats of condensation and of deposition, respectively. If the vapor changes to a liquid before freezing, the latent heat of condensation is first released, followed by the release of the latent heat of freezing (fusion). The heat released in the two-step process is the same as the direct change from vapor to solid, however. This heat release warms the air parcel. In doing so, the buoyancy of the parcel relative to the surrounding air increases, and this contributes to the parcel's further rise. We can see the continued ascent of these parcels in cumulus clouds that reach great vertical growth as they build.

Droplet Growth

Within the cloud, there must be growth of the cloud droplets to sizes that can fall to the ground as rain without evaporating. Cloud droplets can grow larger in three ways: by condensation, by collision and coalescence, and by freezing. The first is by the continued condensation of water vapor into existing cloud droplets, increasing their volume until they become drops. While the first condensation of water onto condensation nuclei to form cloud droplets occurs rather quickly, continued growth of cloud droplets in this manner proceeds very slowly. Growth by collision and coalescence of cloud droplets (and then the collision of raindrops with cloud droplets and other raindrops) is a much quicker process. Turbulent currents in the clouds provide the first collisions between droplets. The combination forms a larger drop that can further collide with other droplets, thus rapidly growing drops in size.

As the drops grow, their fall velocity also increases relative to the fall of surrounding drops, and they can collide with slower falling droplets. A 0.5-millimetre-radius drop falling at a rate of 4 metres (13 ft.) per second can quickly overtake a 0.05-millimetre drop falling at 0.27 metres (0.88 ft.) per second. When drops are too large, however, their collection efficiency for the smallest

Raindrops grow by colliding and combining with other drops, and scavenging smaller drops into their mass.

drops and droplets is not as great as when colliding drops are nearer in size. Small droplets may bounce off or flow around much larger drops and therefore do not coalesce. A drop about 60 percent smaller in diameter is most likely to be collected by a large drop. Clouds with strong updraft areas have the best drop growth because the drops and droplets stay in the cloud longer and have many more collision opportunities.

It may seem odd, but the best conditions for water drop growth occur when ice crystals are present in a cloud. When in small droplet form, liquid water must be cooled well below 0°C (32°F) before freezing. In fact, under optimal conditions, a pure droplet may reach –40°C (–40°F) before freezing. Therefore, there are areas within a cloud where ice crystals and water droplets co-exist. When ice crystals and *supercooled* droplets are near each other, there is a movement of water molecules from the droplet to the crystal. This increases the size of the ice crystal at the expense of the droplet. When the crystals grow at temperatures around –10°C (14°F), they begin to develop arms and branches, the stereotypical snow crystal. Such crystals not only are efficient at growing at the expense of water droplets, they also easily stick to one another, forming large aggregates we call snowflakes.

Precipitation

Finally, the drops grow to a size that can fall to the surface without evaporating in a reasonable time, and we have reached the top rung: precipitation. The following table gives some typical drop diameters for various rain types, using cloud droplets as a reference size:

Drop/Droplet Diameters in Relation to Cloud Droplets

Particle Type	Drop Diameter Range (mm / in.)	Typical Diameter (mm / in.)	Typical Volume Relative to Cloud Droplet
Cloud Droplet	0.01 to 0.02 mm / 0.0004 to 0.0008 in.	0.012 mm / 0.0005 in.	1
Large Cloud Droplet	0.07 to 0.15 mm / 0.003 to 0.006 in.	0.1 mm / 0.004 in.	579
Mist Droplet	0.02 to 0.2 mm / 0.0008 to 0.008 in.	0.5 mm / 0.02 in.	72,300
Drizzle Drop	0.2 to 0.5 mm / 0.008 to 0.02 in.	1.2 mm / 0.05 in.	1,000,000
Raindrop	0.5 to 5.0 mm / 0.02 to 0.2 in.	3.0 mm / 0.12 in.	15,600,000
Large Raindrop	>5.0 mm / >0.2 in.	6.0 mm / 0.24 in.	125,000,000

Of course, not all precipitation falls as rain. A fair amount of the world's precipitation falls as snow or some other solid water form. Common forms are snow, hail, sleet or ice pellets, graupel or soft hail, and snow grains. Actually, outside the tropical regions, it is likely that much of the precipitation begins in solid form and becomes liquid rain only when it melts while falling through air with temperatures above freezing.

One exception to the general process followed by almost all cloud and precipitation formation is the fall of diamond dust, in which extremely cold temperatures in the surface air layer change water vapor directly to snow crystals that flutter and fall slowly to Earth. But with the warming of spring, the occasions when ascent occurs become more frequent and soon our thoughts are on April showers.

Eighteen

Raindrops, So Many Raindrops

As I write this, a light rain is falling over southern Vancouver Island. The drops are barely perceptible in the shallow puddle on the flat rooftop of the building next door. If I were walking out there, I would feel only the gentlest touch from the falling drops. At the same time, across the central plains of the U.S., come reports of torrential rains and flash flood warnings as a line of severe thunderstorms bullies its way eastward.

With all those visions of raindrops dancing in my head, it's a good time to look at the individual drops more closely. When we think of raindrops, we generally have a narrow vision of what they are, how they look, and what their size and composition are. While individually they do not have the delicate beauty of a snow crystal or leave the hard evidence of a hailstone, raindrops hold a spectrum of variability within their watery mass.

Early Experiments with Raindrop Size

A commonly held belief about raindrops is that they are all the same size in a given rainfall. As the table on page 77 showed, a typical drizzle drop is about one half the diameter of the largest raindrops forming in heavy showers. However, although the various precipitation types have a characteristic drop size, there is usually a wide range of drop diameters in rain, of which the typical diameter in the table is the most frequent size. Small drops generally outnumber large drops, but as the intensity of the rainfall increases, the number of larger drops increases. The largest drops are found only in downpours with rainfall rates greater than 5 centimetres (2 in.) per hour.

The first studies of the nature of raindrops began in the late 1800s. Philipp Lenard of Germany was a brilliant experimental physicist who received the Nobel Prize in Physics in 1905 for his work with cathode rays. He studied or taught at many of the major universities in Germany and Eastern Europe during the late nineteenth and early twentieth centuries. Lenard began studying raindrops in 1898. At the same time, investigations of raindrop size were taking place in Europe and the United States. It appears Lenard was not aware that E. J. Lowe (1892) and J. Wiesner (1895) had made the first measurements of raindrop size. Nor was he aware of the work being undertaken by American farmer/scientist Wilson A. Bentley, who we met earlier in the discussion on snowflakes and will meet again shortly. Lenard published the results of his extensive investigation in June 1904 (four months before Bentley published his findings on raindrops) in a paper that presented his work on the shape, size, and stability of raindrops during their descent from clouds.

Faced with the problem of how to measure raindrop sizes during a rainstorm, Lenard used as a drop collector blotter paper dusted with a water-soluble dye. When raindrops fell on the impregnated blotter, they produced colored wet spots that could be measured. Concerned that the size of the wet spot on the blotter paper might not reflect the true size of the drop that made it, Lenard undertook an experiment to establish whether a relationship between the spot size and drop diameter existed. By dropping known-sized drops onto the blotter paper and measuring their splash print, Lenard developed a calibration curve for the method.

Lenard partitioned his raindrop data into 0.5-millimetre-diameter (0.02-in.) intervals, reporting it as the number of raindrops of a particular size range falling on an area of 1 square metre in one second. He collected only ten field samples of drop size distribution using this technique, and therefore could not draw many general conclusions relating drop size to rain event conditions. Since Lenard found no drops with diameters less than 0.5 millimetres (0.02 in.), he concluded that the updrafts in clouds must be of sufficient strength to prevent such small drops from falling out. At the upper range of sizes, he recorded only one drop in the 4.75- to 5.25-millimetre-diameter (0.19- to 0.21-in.) range.

Much of Lenard's subsequent work focused on the behavior of raindrops as they fell from clouds. To do this, he constructed an innovative vertical wind tunnel in which he could vary the upward speed of the airflow to simulate atmospheric updrafts. By adjusting the airflow rate, he could briefly balance a drop in the air stream. This balancing act simulated the aerodynamic forces acting on a drop falling freely through a still air column—and a balancing act it was. The airflow turbulence levels in his wind tunnel were so high that drops could not be held steady for more than a few seconds.

Using the wind tunnel to observe the behavior of drops of known size in an airstream, Lenard determined that small drops up to about 2 millimetres (0.08 in.) in diameter fell as spheres. Larger drops, however, deformed while falling and became unstable at diameters greater than 5.5 millimetres (0.22 in.). They lasted less than a few seconds before breaking apart in the airflow, torn asunder by the turbulent aerodynamic forces buffeting on the drop. This observation, combined with the lack of larger drops in his rainfall measurements, led Lenard to conclude that the maximum drop size possible in nature was no larger than 5 millimetres (0.20 in.).

The first evidence for the wide distribution of raindrop sizes within rainfalls was gleaned from a series of experiments carried out in the northeastern United States by Wilson A. "Snowflake" Bentley. Bentley's apparatus for gathering raindrops was a marvel of simplicity—he collected raindrops in a pan of wheat flour.

Bentley had previously found that after exposing a pan of flour to the rain, dough pellets would form under the flour from the combination of the flour with water from the drop. If he let these pellets harden, he could extract them from the flour and measure them. Through careful experiments using known drop sizes, Bentley determined the pellets to be approximately the same size as the falling drops. Thus, he could determine a size distribution of drops from each exposed pan. To better understand the distribution of various sizes of raindrops within a rainfall event, he divided his raindrop "fossils" from each individual collection into five size categories: very small, small, medium, large, and very large.

The next step was to determine if raindrop sizes or their size distribution varied with rainfall conditions. This required many samples and a record of conditions under which the raindrops fell. Over several years, Bentley amassed 344 separate size distributions sampled from 70 individual storms, including 25 thunderstorms. Unlike Lenard, Bentley kept extremely detailed records of the weather conditions the raindrops were collected under.

Bentley reported the results of his raindrop research in October 1904. He found only a few samples (around 7.5 percent) showed little variation in drop size, being composed mainly of either all small drops or all large drops. He concluded that drops from all size categories were present in most

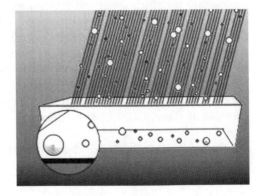

Schematic of the flour pan used by W. A. Bentley to determine raindrop size. Raindrops hitting flour produce "fossil" droplets of flour dough.

rainfalls, although smaller drops greatly outnumbered larger ones. Low-height clouds generally produced few large drops. The largest drops (more than 5 millimetres [0.20 in.]) collected fell from thunderstorms that reached high into the atmosphere. Bentley found the maximum raindrop size to be around 6 to 8 millimetres (0.23–0.31 in.) in diameter, in general agreement with Lenard's findings that drops larger than 5 millimetres (0.20 in.) usually broke apart.

Unfortunately, Bentley was both ahead of his time and not fully appreciated by contemporary scientists, who did not take this innovative, but not formally trained, man seriously. Forty years passed before his work was finally given serious recognition. In 1943, researchers from the U.S. Soil Conservation Service used the flour pellet technique to measure raindrop size as a function of rainfall intensity. These results, combined with the measurements of others, led to the development by Canadian scientists J. S. Marshall and W. M. Palmer of a mathematical relationship between raindrop size distribution and rain intensity. Such relationships have been instrumental in the development of radar as a tool to observe precipitation and determine its accumulation rates.

The Speed of Falling Drops

Philipp Lenard also used his wind tunnel to determine the fall velocity of drops by increasing the flow rate until the drop became suspended. This flow speed was equivalent, he correctly reasoned, to the drop's fall velocity in still air. He found that the fall speed increased with drop diameter until a size of 4.5 millimetres (0.18 in.). For larger drops, however, the fall speed did not increase beyond 8 metres (26 ft.) per second. He attributed this to the changes in drop shape caused by the airflow as the drop size increased. The change in shape increased the drop's air resistance and slowed its fall rate.

How fast does a raindrop fall in the natural environment? Assuming a raindrop falls through still air, two forces acting on the drop determine its terminal (maximum) velocity. First, gravity pulls the raindrop earthward. If the drop fell through an airless environment, it would accelerate at 9.81 metres (32 ft.) per second squared until it hit a surface, its speed on contact depending on the fall distance, or alternatively, the fall time. Thus, a raindrop falling from 500 metres (1,641 ft.) through a vacuum would smash into the ground at nearly 100 metres per second (223 mph). Now that would hurt!

Fortunately, the second force acting on the drop, *aerodynamic resistance* or *drag*, slows the falling drop. When the two forces balance, the drop reaches its terminal velocity. Thereafter, it falls at that terminal speed until it hits the ground. Aerodynamic drag depends on the raindrop's shape, size, and speed. Drag quickly brings the falling drop to its terminal velocity, assuming no influence from up- or downdrafts during its fall. A drop 2 millimetres (0.08 in.) in diameter reaches terminal velocity after about 2 metres (6.6 ft.) of descent. A typical drizzle drop falls at 2 metres per second (4.5 mph). A raindrop 2 millimetres (0.08 in.) in diameter falls at 6.5 metres per second (14 mph), while a large raindrop measuring 5 millimetres (0.20 in.) across—the size of a small housefly—falls at around 9 metres per second (20 mph). For comparison, a baseball dropping from a towering pop-up hits the outfielder's glove at a terminal speed of 42.8 metres per second (95 mph).

Since the drops are falling within an environment where updrafts and downdrafts are common, they will fall at speeds different from those calculated above. A drop's falling speed in such an environment can be calculated by adding the speed of the up/downdrafts to its terminal velocity, where the updraft is considered a negative speed. A drop falling at a terminal speed of 6.5 metres per second (14 mph) caught in a 2-metre-per-second (4.5-mph) updraft, would fall relative to the ground at 4.5 metres per second (10 mph). If the updraft speed is greater than the terminal speed, the drop will rise rather than fall. That same drop falling in a downdraft of 2 metres per second (4.5 mph) descends at 8.5 metres per second (19 mph).

Purity of Raindrops

A common misconception holds that raindrops are pure water. The fact is, rain would not fall without some degree of impurity in the air to act as a seed. This seed may be a chemical salt, an acid droplet, a speck of dust or soil, even a bacterium. Cloud droplets or ice crystals, which precede raindrop formation, require such seeds to readily form from the water vapor in the air. Some seed materials are better than others in forming droplets or crystals. When the drops formed from these cloud droplets or ice crystals fall earthward, they carry their seeds with them.

To get an idea of the amount of seed material required to form a raindrop, let's look at a simple example. Near the ocean surface, the air is filled with billions of minute salt particles (diameters between 0.5 and 10 micrometres) floating above the water. A typical rain cloud forming above the ocean waters contains billions of cloud droplets, each one likely formed on at least one sea salt particle. It takes a million cloud droplets to provide the water contained in one small-sized raindrop, so we can assume that each raindrop contains the material from a million or more salt particles. The same can be said of a raindrop formed from any other seed type or combination of seed types.

In addition, as raindrops fall, they may collect chemicals, particles, bacteria, pollen, and other seeds from the air they pass through. This process is known as *scavenging*. How well a given raindrop scavenges material from the air around it depends on a number of factors related to both the drop and the potentially scavenged material. Scavenged particles are usually deposited onto the surface the drop lands on. Scavenging can be a beneficial process when it cleans the air of "impurities," but the eventual deposition of these impurities on a surface may not always be welcome.

In the last twenty-five years, *acid rain* has been gaining recognition as a major environmental problem. "Pure" rainfall is usually slightly acidic, with a pH of about 5.8, due to the presence of carbon dioxide gas in the drop. Acid rain, or more correctly *acidic precipitation*, forms when cloud droplets and raindrops scavenge or use as seeds acidic materials released as air pollutants, such as sulfur dioxide. When deposited on surfaces such as buildings, snowfields, and lake surfaces, the acidity accumulates, eventually becoming a hazard to the health of plants and animals and damaging materials, particularly marble, limestone, and iron.

Raindrops may also collect dust and sand particles or pollen grains in sufficient quantity to give the resulting rain a distinct color. Reports of storms "raining blood" usually come from rainstorms that have scavenged red dust or sand from air blowing off desert areas such as the Sahara and Australian interior deserts. Many folk legends and myths of omens falling from the sky likely originate from the inclusion of such materials in rainfall.

Raindrop Shape: No More Tears

The time has come to shatter a long and tightly held belief: raindrops are *not* tear-shaped.

Ask anyone to imagine a raindrop, and odds are they will picture a teardrop shape. Literature, poetry, and song are filled with allusions to raindrops as tears from the sky or from heaven. The analogy likely began when someone noticed that when raindrops hit a surface, they roll off like tears flowing down the cheek, having a rounded front edge with a pointed tail.

Raindrops are not shaped like tears.

In truth, every airborne raindrop is spherical in shape as it begins to fall. Then, unless it remains rather small, the drop's shape changes to something more like a hamburger bun. The distortion is caused by the air flowing around the drop, which pushes against the drop's lower surface, flattening the base as it falls. This aerodynamic drag force can further deform the largest drops into a sagging dumbbell shape, eventually causing the biggest ones to split into two or more smaller drops. (If a drop fell through a vacuum, it would retain its spherical shape.)

Even the hamburger-bun shape—based upon observations of single drops in a steady airflow—is idealized. When rain falls, its drops have many different sizes, and each drop size falls at a slightly different speed. The smallest drops may not fall at all, being suspended or perhaps forced upward by ascending currents of air until they grow large enough to overcome the resisting updraft force. As a result, there are many collisions between raindrops. Some collisions cause distortions in the drops' shapes as they bounce off one another. Others cause drops to coalesce, forming a large drop in the process. Some collisions cause drops to break apart.

If we could isolate a moderately sized drop in a rainstorm and

True raindrop shape is determined by wind flow patterns and aerodynamic drag forces.

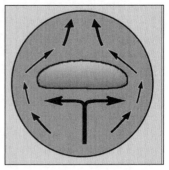

follow it from formation to splashdown, we would see not a teardrop or a sphere, but an ever-changing, quasi-spherical shape. As it falls, our drop may grow in size by collecting other drops or perhaps split apart due to collisions or by attaining an unstable size.

Despite the fact raindrops are not teardrop-shaped, we continue—yours truly included—to depict them in that familiar shape. Perhaps the streamlined shape of the teardrop implies falling motion more than the true shape does? Although I know better, it is likely I will continue to use artistic license when depicting raindrops for just this reason.

Twenty

Up, Up, and Away

During the late 1960s, the Fifth Dimension produced a hit recording titled "Up, Up, and Away." For those unfamiliar with the song, it is about riding in a balloon and floating among the stars over a peaceful Earth below. I have often considered it one of my theme songs as I sit watching the weather unfold around me.

This afternoon, I see a flock of small cumuli gathering like sheep over the Gulf Islands and the Coast Range on the continent to the east, moving "up, up, and away" from me, riding the winds eastward. The scene engendered memories of sitting on a hillside one long-ago April, watching cumulus clouds spring up over the flat terrain of Michigan. What I was supposed to be doing that spring day was studying for my meteorology finals. What I was doing instead was studying the weather—field research of the best kind, watching cumuli puff into bubbly towers.

Just as I was back then, I am watching the visible manifestations of *updrafts*, rising currents of air. The updrafts I observed in Michigan resulted solely from the daytime heating of the ground, but here on Vancouver Island, terrain effects are also hard at work.

A general definition for updraft is a rising current of air, although the term is often limited to upward-moving currents within large cumulus clouds and thunderstorms. Storm updrafts are the athletic big brothers of the updraft family, reaching speeds as swift as 80 km/h (50 mph). Most of the family, in contrast, are tiny upward currents that rarely exceed 10 centimetres per second, or 0.36 km/h (0.22 mph).

We can identify at least five main driving forces for updraft formation in the atmosphere. Depending upon the time and space scales of interest to the observer, these forces may work alone or in concert; in some cases, one type may initiate another.

Turbulence is the chaotic twists and turns, ups and downs, working on a very small scale (microscale) in a fluid flow such as the wind. Turbulent flow is the opposite of smooth (*laminar*) flow.

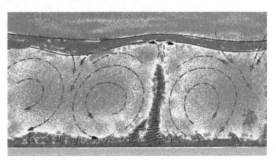

Thermal heating of the surface generates heated plumes (dark regions) that rise from the ground while cooler air descends to replace them.

Topographical forcing is the disrupting effect on airflow from the kind of terrain variations seen on a topographical relief map, such as hills and valleys, mountains and basins, and canyons and shorelines. Vegetation and human structures may also produce smaller-scale topographical forcing.

Air density forcing, commonly called *buoyancy*, produces updrafts due to the density difference between a lighter air parcel and its environment. Most often, the density difference is due to the temperature differential between the parcel and its surroundings. Warmer air is lighter air and will rise through cooler air. The water vapor content of an air parcel also affects its density. All other factors being the same, a volume of air with greater water vapor content will be lighter than a similar, drier volume and thus rise.

Convergence is the process whereby fluid streams move together into a smaller area. In the lower atmosphere, if winds converge at the surface over a relatively small area, the inward push of the air-molecule crowd forces those in the central area to begin an upward air current, an updraft.

Divergence is the opposite flow situation to convergence—fluid streams flowing apart from a common area. High in the atmosphere, where the physical barrier of Earth's solid surface is too far below to be effective, a region of upper-level divergence can establish an updraft below it as air is drawn up to replace that leaving the divergence zone.

Another updraft mechanism is *frontal lifting*, the lifting of air along warm and cold frontal zones. I consider this a second-rank forcing because updrafts result from the combination of at least two main driving forces, air density and convergence forcing (at times enhanced by divergence aloft). The process of frontal lifting works like this. When warm and cold air move toward each other—converge—to form a front, the lighter warm air rises through buoyancy forces above the colder air. Generally, the greater the temperature, the faster the updrafts formed at the front. A region of divergence above the front can enhance the updrafts.

Watching Updrafts

I am most interested in weather phenomena occurring on time and space scales that we can personally sense: the small end of the regional scale, the *mesoscale*, and the microscale. On these scales, both updraft forcing by topography and by buoyancy may reveal visible signs of updrafts that the astute observer can easily recognize once familiar with them.

Since I am surrounded by rugged mountainous terrain rising from sea level on Vancouver Island, I see topographical influences on updrafts more often than if I were a resident of flat terrain areas such as Saskatchewan or Florida, where buoyancy influences are the most dominant. Mountains or high hills are good places to watch for updrafts.

Where topography presents a mostly impenetrable barrier to the airflow, it forces the air to rise over the obstacle, producing updrafts of various strengths. However, any vertical step-up formed by a topographical element, even one as small as a tree or bush, can induce updrafts unless other atmospheric conditions work in opposition, or the breadth of the obstacle (such as a telephone pole) makes flow around it easier than flow over it. All else being equal, the sharper the vertical change in terrain height presented to the flow, the faster the updraft produced by it.

Clouds are the best visible indicators of updrafts formed by terrain. Such *orographic* (terrain-produced) clouds form when moist air, ascending to cross a terrain barrier, reaches its condensation level. Above this level, some of its water vapor will condense into liquid droplets or deposit as ice crystals and form visible clouds. Clouds that form a cap or band across the summit region, or that extend outward from the summit region, are called *cap* and *banner* clouds, respectively. They are members of the *cumulus* family as they are formed by convective updrafts, being *altocumulus* if formed high in the atmosphere or *cirrocumulus* if formed even higher as ice clouds. In certain situations, cap clouds resemble lens-shaped bodies and are known as *lenticular clouds*. Lenticulars can be so unusual in appearance that they have been reported as UFOs.

Larger thermal plumes form cumulus clouds and initiate updrafts in vertically developed cumuli.

To the casual observer, these clouds may appear to remain fixed over the summit even when a strong wind is blowing. Concentrated viewing or time-lapse cinematography, however, will show cap clouds to be fluctuating masses, changing their shape and size from one instant to the next. What is actually seen is a train of individual air parcels rising on the windward side above their condensation level, then again dropping below the condensation level on the leeside of the summit ridge. Once below the condensation level, the liquid cloud droplets evaporate back into invisible vapor, the clouds rapidly dissipate, and the air clears.

An interesting variation on summit banner clouds occurs when the terrain obstacle induces a vertical oscillation, or *standing wave*, in the crossing airstream that persists for some distance downwind. The standing wave pattern that develops causes the air parcels streaming downwind to oscillate above, then below their condensation level. The wave motion forms stationary bands of thin cumulus-type clouds separated by clear air bands extending downwind from the ridge. These cloud bands are called *wave clouds*.

Often, airflow over mountainous terrain produces strong updrafts that lift the air sufficiently to allow the clouds to grow into

Cumulonimbus rapid growth forms a mushroom cloud as it becomes a thunderstorm.
(NATIONAL OCEANIC AND ATMOSPHERIC ADMINISTRATION/U.S. DEPARTMENT OF COMMERCE,
NOAA HISTORIC NWS COLLECTION)

precipitating cumulus towers or thundering cumulonimbus. The added uplift provided by coastal mountains produces significant rain or snowfalls or enhances that from passing storm systems such as those that strike the Pacific Northwest from southeastern Alaska, through British Columbia, and into Washington and Oregon. The additional contribution of orographic precipitation provides ideal moisture conditions for the trees and vegetation that make up the region's great temperate rainforests to flourish.

While clouds provide long-lasting, large indicators of topographically induced updrafts, some animals and plants can also make regions of updrafts apparent. The swaying of tree branches or the flight of their lost leaves can reveal vertical currents. If released in great quantities, seeds from species such as milkweed and dandelion and pollens of other plant species can also mark the presence of updrafts, but usually their small size and low density make following their windblown travels difficult except on the smallest scales. Small animals, particularly spiders, mites, and insects, hitch rides on the wind to extend their dispersal by updrafts to launch themselves into the airstream, but they too require close inspection to see. For my money, the champions at exploiting updrafts are members of the bird family. Among the most expert bird species are the great soaring raptors—eagles, hawks, and vultures—but many sea birds are equally adept and can remain aloft for very long periods of time.

Today, I watched gulls and other sea birds use the topographically induced updrafts along the shore of the Gulf Islands to gain a higher perspective on the sea field below. They can also exploit the air currents above the ocean waves to glide nearly effortlessly just above the water surface for long distances. They catch the small updrafts on the windward wave face, which catapult them to a higher altitude. Gliding slowly downward under the pull of gravity, they seek another vertical push just before reaching the wave surface.

Smoke from industrial operations or surface fires may also reveal topographic updrafts, but care should be taken in interpretation as the heat that produces the smoke imparts added buoyancy to the smoky parcel. Thus the rise is often due to a combination of topographical and buoyancy forces.

Small-Scale Updrafts

I could go on for pages about other indications of updraft presence, but I will leave you to explore them on your own. I finish with a small-scale viewing delight, particularly if you share it with your kids. (If you have none to awaken the child within you, go borrow a few from family or friends for a couple hours.) Small-scale updrafts, as well as the other twists and turns in the wind field around us, can be seen around buildings, trees, fences, hedges, and other structures using a simple tracer: soap bubbles.

First, choose a day with a comfortable temperature and light to slightly moderate winds. Too light a wind and you will see mostly buoyancy-derived currents (not a bad thing). Too strong a wind and the bubbles will disappear too quickly.

Next, get some equipment. There are many varieties of bubble makers available; the choice is yours, but a bowl of dish soap solution and a small frame bent from a paperclip or short length of wire is really all that's needed. The best bubbles for watching the wind are half again larger than a quarter. Bigger ones are usually too fragile; smaller ones disappear from view too quickly.

Finally, start bubbling! Oops, did you remember to round up the kids to play with you?

Setting off a few bubbles at a time allows me to pick one and focus on that bubble's trajectory. However, setting off a small bubble armada can reveal details of the wind's spatial variability. With the aid of kids, you can launch bubbles from various locations and see how the currents vary around you.

While you are playing/experimenting, don't forget to give an explanation or two to your bubble crew, even if it is only to get them involved: "Did you see the way that bubble rose and twisted?" A similar play experiment, on a day when the wind is calm or nearly so and solar heating is strong, can reveal details of local buoyancy-driven updrafts.

Twenty-one

Mirages: Not Just for Deserts Anymore

Driving down the highway on a hot day, I look at the pavement ahead of me and see it covered with what appears to be a pool of water. But I never reach that wet pavement because, rather than a pool of water, what I observed is something commonly known as a *highway mirage*. Years before highways spiderwebbed the countryside and huge areas of land were paved or built upon, people would have called such a sight a *desert mirage*, or simply a *mirage*.

For many the word "mirage" brings thoughts of thirsty travelers moving slowly across desert sands toward the image of a pool of clear water. Whether the image is of cowboys in the American or Mexican desert, or of French Legionnaires lost in the Sahara, we likely consider the mirage to be an illusion of an overstressed mind, or a figment of the imagination of someone parched with thirst. But a mirage is a real image and it can be photographed.

The desert mirage and the highway mirage represent a common form known as the *inferior mirage*. Another mirage form is the *superior mirage*, which is most often viewed at the times of sunrise or sunset or over large bodies of cold water or snow or ice fields.

A mirage is defined as "an optical illusion due to atmospheric conditions by which reflected images of distant objects are seen." I'll excuse the lexicographers on the use of "reflected" because it seems more appropriate, even if incorrect. Actually, they were only two letters off; the proper technical term should be "refracted."

The illusional part of a mirage emerges from the way our mind processes what our eyes see. When we view the light rays from an object, our mind makes the usually correct assumption that a straight line from our eyes connects directly to the location of the object, or any detail on it. It has no way to correct for any possible bending of the light. (Although, through reasoning, we can mentally make this correction after the fact.) For example, when we see a pool of water on the ground formed by inferior mirage conditions, we expect that "pool" to actually be on the ground.

All mirages form when light passing through the atmosphere is refracted by the differences in the density of the air. The *index of refraction* (a measure of the degree of bending of the light by the medium) for air is directly proportional to the air density, and air density is proportional to its pressure (density increases as pressure increases), and inversely proportional to its temperature (density decreases as temperature increases).

The effect on the setting/rising sun's position derives from the increase in density due to atmospheric pressure increases as the solar rays come from the near void of outer space toward the earth's surface. In the lowest atmospheric layers, however, density differences are more dependent on the temperature changes than the pressure changes. Therefore, when the temperature varies significantly over a short height (known as the *temperature gradient*) in the lower atmosphere, the path of light rays traversing that layer can be bent from a straight line path. The bend is always toward the direction of the higher density (and therefore lower temperature). Some very interesting optical effects can occur when the atmosphere has multiple layers of differing density, such as the appearance of boats floating upside down in midair.

Superior Mirages

One form of superior mirage is so common that even many scientists do not recognize it initially as a mirage. Have you ever seen the sun lying right on the horizon? What was the actual position of the sun at that time? Yes, this *is* a trick question.

The fact is, anytime you *see*, from the ground, the full solar disk directly on the horizon, all or part of it is actually below the horizon. The bending (*refracting*) of the incoming solar rays by the atmosphere causes the optical illusion that the sun is actually on the horizon. This happens daily at both sunrise and sunset, and adds extra minutes of daytime to each day.

When the air at Earth's surface is very cold relative to that lying above it, you may see a

Warm air layer over cool air layer distorts light path, giving the vision of a floating, inverted sailboat.

superior mirage. "Superior" indicates that the image we see is *above* the actual image position. On the opposite side, when very warm air underlies relatively cool air, you should look for "inferior" mirages, so called not because of the quality of the image but because the image we see is *below* the actual position of the object.

For a superior mirage to occur, the air close to the surface must be much colder than the air above it. Such conditions are common over snow, ice, and cold water surfaces. When very cold air lies below warm air (called a temperature inversion), light rays are bent downward toward Earth's surface, thus tricking our eyes into thinking an object is located higher or is taller than it actually is. As a result, the superior mirage makes objects appear to be floating in the air or causes objects actually located below the horizon to appear above it (remember the setting sun example), a condition called *looming*. The superior mirage can also stretch the image of an object, making it appear to be taller than it actually is, a condition called *towering*.

The superior mirage may cause the image (or parts thereof) of an object to appear:

- visible, even though it is actually located below the geometric horizon (*looming*);
- lifted well above its actual position (*lofting*);
- inverted from its normal image;
- multiplied and either upright or inverted;
- taller, larger, or closer than it actually is (*towering*); or
- shorter, smaller, or farther away than it actually is (*stooping*).

Perhaps the most common superior mirage is a looming one. The bending of the light ray paths under looming conditions increases as the temperature inversion increases in the air layer. Thus, the greater and deeper the inversion, the higher the object appears in the sky because our eyes interpret the bent light path as straight. This is illustrated in the accompanying diagram. Because of this bending, we see the object floating in the sky, above or even attached to the original object. In cases of strong looming, the image may appear very high in the sky. Looming of a distant ship may be the source of the many legends of flying ghost ships seen by mariners over the centuries. Since the actual ship may be below the geometric horizon, sailors or shoreline viewers might never see it except as a mirage image.

The weather conditions that form superior mirages are quite common at night when radiational cooling of the air near Earth's surface, under clear skies and light winds, is strong. Mirages at this time are seldom noticed, however, since it is usually dark and most of us are indoors. But artificial lighting, particularly from buildings

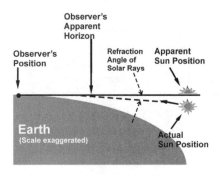

Superior mirage formed by atmospheric refraction makes the sun visible when it actually sits below the horizon.

and vehicle headlights, can be *lofted* (appearing to come from well above the surface) from their expected position. As a result, the superior mirage could be the source for some UFO sightings. Here's why. The light from headlights on automobiles moving up a sloped road can be refracted under proper atmospheric conditions so that they appear to come from the heavens rather than from the road surface. They can then move quickly across the sky or disappear suddenly when the vertical angle of the light beam on the moving vehicle changes.

Robert Greenler, in his book *Rainbows, Haloes and Glories*, reported on one interesting superior mirage viewing on the night of 26 April 1977. When residents of Grand Haven, Michigan, looked westward that night across the relatively cold waters of Lake Michigan, they distinctly saw city lights and a flashing red beacon. But the nearest urban area to the west was Milwaukee, Wisconsin, 120 kilometres (75 mi.) away, well below the horizon and normally not visible. When a Grand Haven resident timed the blink rate of the flashing red light and linked it to the Milwaukee Harbor entrance beacon, their sightings were confirmed to have indeed been Milwaukee. U.S. Weather Service records also confirm that strong inversion conditions were present that night. The unseeable had briefly become visible.

Superior mirages can also have a playful side, inverting images of ships at sea and hanging them in midair or constructing dancing fairy castles over the waters, known as the *Fata Morgana*. There are even recorded instances when the polar night was broken several days prior to the expected polar dawn. This effect, called the *Novaya Zemlya*, after the Arctic island of the same name, was first reported in 1596, when an ice-bound expedition searching for the Northeast Passage to the Orient surprisingly watched the sun rise two weeks earlier than expected and then set again!

Arctic Mirages

Don't feel bad about being tricked by a superior mirage. Several Arctic explorers in search of the Northwest Passage were convinced huge ice mountains blocked their path. For example, in 1818, Sir John Ross observed and named the Crocker Mountains in the

Canadian Arctic, which he estimated to be 50 kilometres (30 mi.) away from his position in Lancaster Sound and blocking his way to finding the North-West Passage. No such mountains were ever found. Professor W. H. Hobbs hypothesized that Ross had actually seen the snowcapped heights of North Somerset Island looming up from their actual position 320 kilometres (200 mi.) distant.

There is ample evidence to suggest that Viking explorers may have discovered new lands because they saw mountains looming above the western horizon, although in most instances they were really only seeing pack ice or small icebergs altered by mirage conditions to appear as mountains. Some historians believe that the voyages of the Viking Erik the Red to Greenland and his son Leif Eriksson to North American shores were realized because they had seen great lands and mountains to the west, visions made possible by persistent and strong superior mirages.

Here is what occurs under these extreme superior mirage conditions, also known as *arctic mirages*. When the temperature of the lower atmosphere increases with altitude at a rate of 11.2 C° per 100 metres (6.0 F° per 100 ft.), the refractive capacity of the air is great enough to cause the path of light rays to bend in an arc equal to the curvature of Earth. This curvature can present an observer with the image of a flat horizon receding to infinity. A temperature gradient greater than 11.2 C° per 100 metres (6.0 F° per 100 ft.) causes the bending of the light ray paths to exceed the curvature of Earth; thus the horizon appears to be raised upward, giving Earth's surface a saucer-shaped appearance. Under this condition, images of objects located at or below the normal optical horizon, such as mountains, glaciers, cliffs, or sea ice rise (loom) into the field of vision, overcoming the normal visual restrictions of Earth's curvature.

The normal viewing distance at Earth's surface depends upon the height of the object being observed and the height of the observer. Disregarding atmospheric effects on light rays, Earth's curvature restricts the distance one can see from the surface. For example, a beach or small iceberg rising 3 to 3.7 metres (10–12 ft.) above the sea surface can be seen from the surface at a distance of no more than 19.2 kilometres (12 mi.) through a clear, normal atmosphere. A mountain peak of 914 metres (3,000 ft.) would disappear at 115 kilometres (72 mi.) distant, one that is 1,520 metres (5,000 ft.) tall at 150 kilometres (94 mi.).

The maximum viewing distance under arctic mirage conditions, on the other hand, is limited only by the light absorption of the atmosphere. Near sea level, the transmission of light is generally of sufficient quality to enable the naked eye to potentially see objects at a distance of up to 400 kilometres (250 mi.) in clear, clean air. However, when the refracting layer is at the upper boundary of a very deep cold layer, the thinner air may permit more light to be transmitted, making visibility in excess of 400 kilometres (250 mi.) possible.

Under arctic mirage conditions, instances of atmospheric visibility extending 320 kilometres (200 mi.) have been reported. In 1937 and 1939, Hobbs documented several occasions during which objects were sighted at distances well in excess of those possible under normal viewing conditions. One significant arctic-mirage sighting occurred on 24 May 1909, when Commander Donald B. MacMillan observed and clearly recognized Capes Joseph Henry and Hekla in Grant Land from his position on Cape Washington on the north Greenland coast 320 kilometres (200 mi.) away.

Inferior Mirages

Inferior mirages commonly form when the air near the ground is much warmer than that above it—the hotter the air, the greater the effect. But it is the temperature *difference* between the altitudes or layers that is a more important factor for mirage formation than the absolute temperature. Highway mirages are as common over dry pavement on sunny winter days as during summer months.

As shown in the accompanying diagram, a layer of very hot air near the ground refracts the light from the sky nearly into a U-shaped bend (a *parabolic* curve). The cause of the mirage illusion is our mind initially interpreting the light rays reaching our eyes as having come along a straight path, which in this example must have originated on the ground. Thus, we see that patch of sky and cloud on the ground and interpret the image as a surface pool of water.

A rule of thumb for inferior mirage formation is presented by Marcel Minneart in his classic book *The Nature of Light and Color in the Open Air*, in which he measures the temperature drop between air at Earth's surface and 1 metre above:

- If the temperature drop over 1 metre is less than 1.7 C° cooler, no mirage forms.
- If the temperature drop over 1 metre is around 2.8 C° cooler, a moderate mirage forms.
- If the temperature drop over 1 metre is greater than 4.4 C° cooler, a strong mirage forms.

To give you some idea of how much hotter the air can be above a paved road in the full sun, temperatures 11 to 17 C° (19.8 to 30.6 F°) hotter at the surface than those measured 1 centimetre (0.4 in.) above the surface have been recorded.

A patch of blue sky mixed with some cloud is the most common vision in an inferior mirage, but any object can be seen that is beyond the mirage's apparent location. If you look closely, you may

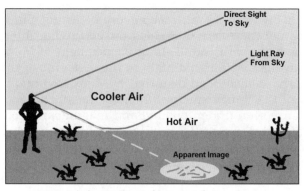

Hot air below cooler air forms an inferior mirage.

see details such as vehicles in the highway mirage. When the temperature gradient is constant through the air layer that the light-ray paths have traversed from the object to the eye, the image will not be distorted. However, if the temperature gradient changes within the layer, either continuously or divided into several distinct layers, the image can be distorted in any number of ways. As with superior mirages, the image may appear taller or stretched (towering) or shorter or compressed (stooping).

One of the more interesting visions that may result from an inferior mirage is the illusion of someone walking on water. Under inferior mirage conditions, a flat surface stretching out ahead of the viewer can appear as if the viewer were standing atop an inverted bowl. This brings the optical horizon closer to the observer than the true geometric horizon of Earth's curvature. With a little distortion added to the picture, it would be easy to mistake a person walking along the shoreline to appear to be walking on the water's surface.

Inferior mirages can be seen anywhere when a surface is hotter than the air above it. They can form over the hot metal of a vehicle, artificial turf, building roofs, even rocks above ice and snow surfaces. The source of the heat need not be the sun, either. A hot engine or barbecue can raise the temperature of its metal hood enough to create conditions for seeing an inferior mirage. Inferior mirages may also form over heated vertical surfaces such as a rock face or wall, giving a surreal view of the surroundings appearing over this surface, including the pool of water illusion clinging to a vertical surface.

Mirage Awareness

Mirages open a whole new aspect to sky watching. Perhaps we have been viewing instances of superior and inferior mirages for years without realizing what we were seeing, or not seeing. Since they were unknown to us, they were often also unseen. What interesting stories may have arisen if the flashing harbor light of Milwaukee had been the only light visible that night to the residents of Grand Haven and had perhaps also been distorted by *Fata Morgana* conditions? What tales of UFOs may emerge when car headlights located below the horizon are seen under looming and magnification conditions of a strong superior mirage? Could those luminous disks darting around the heavens at unusual speed and motion be only the beams from the headlights of a Volkswagen Beetle bent through the air?

Now that you know what to look for, take some time to look closely when conditions are right for mirage formation, and you may observe some surprising scenes. Next time you are out for a drive or a walk and see that pool of water in the distance, take some time to look more closely. You may see all kinds of detail in the mirage that you have never seen before. Perhaps you might get a feel for the stuff legends are made of.

April Showers

April showers bring May flowers, so the old saying goes. On this April day, the wrens and house sparrows were singing their territorial songs. The morning dawned, fittingly, a robin's egg blue with just a hint of a breeze, but by mid-afternoon a cauliflower horizon moved overhead and the temper of the sky changed. No longer a lovely blue, the sky's color dulled to shades of gray and blue-gray and dirty white.

Then I heard it. At first it was a *tip-tap* on my bicycle tarp, then a *pitter-pat*, and finally, as the day drew darkly around me, a steady *whoosh* of sky tears. For ten to fifteen minutes, buildings and trees in the distance disappeared behind the aqueous curtain. Then, as quickly as it had exited, the sun's light burst forth, reflecting in multicolored hues from the oil-water mélange on the pavement below. It was an April shower, a common form of precipitation here among the mountains of the Pacific coast. Showers are also very common during April across most of North America and can often be snow showers as well as rain showers, particularly in Canada and the northern tier of the United States.

Showers are a distinct form of precipitation, whether rain or snow. The accepted technical definition for *shower* is precipitation falling from a convective cloud, characterized by the suddenness with which it starts and stops, by rapid changes of intensity, and usually rapid changes in the appearance of the sky. The key words here are "falling from a convective cloud" and "suddenness." Showers are not always associated with those large-scale storms that dominate winter weather, when precipitation usually falls steadily from *stratiform* clouds surrounding the low-pressure cell. However, in areas such as the Great Lakes basin, much snow may fall as snow showers or snow squalls in the cold air before or after the storm system when convective clouds boil up as the cold air crosses relatively warm waters.

April showers are more than just a poetic choice of words to rhyme with "flowers." April is a transitional month in much of North America in the character of precipitation—winter's vigorous cyclonic activity is on the wane, and steady frontal rains are becoming less frequent. From now until late in the autumn, much of the rain falling across the continent will come from convective clouds.

In April, the upper atmosphere still retains much of winter's cold temperatures, but below, Earth's surface is nearly bare of snow and likely very moist. This allows the sun's growing heat to evaporate moisture from the soil and also significantly warm the air adjacent to the surface. Warming air below and cold air above results in the rapid rise of warm bubbles of moist air, each bubble generally capped with a cumulus cloud as soon as it reaches the condensation level.

If conditions are right, these bubbles grow into larger clouds, which form small raindrops in the ascending air. When the updrafts are strong, the drops will grow to larger sizes. Eventually, either the updraft collapses, when the air is no longer sufficiently warm to continue rising, or the drops become too heavy for the updraft to support, and they fall to Earth as rain. Because the updraft collapse can be sudden, the onset of rain is also sudden, giving the precipitation its showery character.

When moist air is forced over a physical barrier such as mountains or large hills, updrafts are initiated. Once the air reaches the condensation level, liquid water will collect in the clouds. After the clouds have been forced over the barrier summits, the supporting updrafts are cut, and as the air descends, some of the liquid falls as showers in the lee of the barrier, the remainder evaporates. Such is the character of much of the rain that falls in the Pacific Northwest in spring.

April showers are generally less severe than summer showers. In high summer, the warm ground quickly heats and forms very large warm air bubbles that are capable of reaching high into the atmosphere. The product of the larger convective cells is usually a thundershower or thunderstorm. Indeed, the added warmth of the air allows summer showers to be much heavier than those in April because of the air's greater water vapor content. For example, if we take cool, saturated air at 6°C (43°F) and cool it to 1°C (34°F), approximately 2 grams of liquid water will condense out per cubic metre. However, if that saturated air was at 21°C (70°F) and cooled by the same amount to 16°C (61°F), 5 grams of liquid water will condense out per cubic metre—more than twice as much to potentially fall to the surface as rain than in the cooler air.

Thus, April showers tend to be gentler falls of rain than their June or July counterparts—and what better way to promote the

emergence of flowers of May. Seed and seedling are caressed by the life-giving rain, not washed out by strong deluge. Given proper growing conditions, the seeds will set and mature, ready to blossom in May and later months.

An April shower. (KEITH C. HEIDORN)

Twenty-three

Why Blue Skies? Why Red?

Two of the more common questions meteorologists hear are: "Why is the sky blue?" and "Why is the sky red at sunset?"

The answer begins where sunlight begins, on our own star, the sun. The fiery nuclear furnaces cause the sun to burn with a heat of around 6,000°C (10,800°F). All objects shine with radiant energy across many different wavelengths depending on their temperature. One of the basic laws of physics states that the radiant energy of any object has a characteristic wavelength in the electromagnetic radiation spectrum dependent on its radiating temperature. For the sun, the peak region occurs in the so-called visible wavelengths (since they are visible to our eyes) between 400 and 700 nanometres that contain the colors from violet to red. You and I also radiate, but do so in the infrared (wavelengths lower than the red visible wavelengths) that are invisible to our eyes, though we shine with a light seen by other animals, such as mosquitoes. We *feel* the infrared, however, as heat.

After traversing the rather empty trail from the solar surface through outer space to our atmosphere, sunlight encounters the denser atoms and molecules of air and assorted dust particles suspended within it. During the split second it takes for the light to penetrate from outer space to the surface and our eyes, these atmospheric constituents have affected the incoming rays. A portion of the beam is reflected right back into space, while some rays are absorbed and re-emitted either back to space or toward Earth's surface. Others are bumped off path, *scattered* in physicists' jargon, and continue their downward journey on a different angle than the primary solar beam.

Within the nature of the scattered portion of the solar beam, we find the answer to our questions about sky color. Scattering treats the various color wavelengths of the *white light* solar beam (containing all the spectral colors) differently, and the effect depends on the size of the matter doing the scattering. Air atoms and molecules scatter the shorter wavelengths, the violets and blues, more readily than the longer visible wavelengths, the yellows and reds. In a dry, clean atmosphere, the scattering of the blue-end colors is about three times that of the red end. The light rays scatter in all directions, and a portion continues downward on a course off the direct solar beam. As these scattered rays continue, they are further scattered, so most light rays reach our eyes coming from directions away from the direct solar beam. These scattered wavelengths are called *skylight*.

Meteorologists call the volume of atmosphere that solar light rays traverse the *optical air mass*. In the cleanest, driest optical air mass, the scattered solar light is predominantly in the blue range of colors, though it does contain some wavelengths of all colors. As a result, we see a blue sky rather than the deep black sky seen by the Apollo astronauts on the airless moon. The more air molecules between the observer and the incident solar beam, the thicker the optical air mass and, therefore, the more light scattering that is possible. For that reason, when we ascend a high mountain or fly to high altitudes in an aircraft, the sky is blacker than at Earth's surface because fewer air molecules to scatter the light around are between us and the beam. The direct beam is also brighter at higher altitudes due to less scatter.

For similar reasons, the angle of the sun from the vertical as it traverses the heavens from dawn to dusk also becomes a factor in determining scattering. When the sun is directly overhead, the amount of atmosphere it must traverse is at a relative minimum. But as it approaches the horizon, that optical air mass increases dramatically, reaching thirty-eight times longer when the sun is on the horizon. We see the impact of increasing optical air mass depth by contrasting the sky color around the sun at midday with that at dusk or dawn. When the sun is near the horizon, the greater depth has scattered away nearly all the blue coloration, and the sun turns from yellow to red, as does the surrounding sky, giving us the fiery reddish sunrise and sunset skies that can be enhanced by some of that light being reflected off cloud surfaces.

Scattering is not the only process affecting the solar beam. The light rays may also be absorbed by the atoms, molecules, and dusts they strike. In absorbing light in specific wavelengths, these constituents, including water in its liquid, solid, or vapor states; dusts, aerosols, and particles; and certain gas molecules usually associated with air pollution, can change the atmosphere's color. In many cases, these changes arise from the greater sizes of these constituents, acting alone or in combination. Water vapor molecules are smaller than the nitrogen and oxygen molecules that make up over 98 percent of the atmosphere, but they have a great affinity for other constituents. For example, water vapor and sulfur dioxide may combine, forming sulfuric acid aerosols—larger molecules that scatter the longer wavelength colors. As a result, the sum of the scattered wavelengths in the full complement of scattering entities produces alternate hues to the dominant blues, ranging from murky grays and browns to vivid golds, oranges, and reds.

In the past, dust and gases hurled into the stratosphere by violent volcanic explosions such as Krakatoa and Mount Pinatubo colored the global skies for several years. Perhaps some of the myths and beliefs related to blood-red skies grew out of the effects of volcanic dusts on sky color. An atmosphere containing a lot of dust would be reddish because the scattering and absorption of the shorter wavelengths would leave only the red in the direct and scattered beams. On Mars, as seen in mission photographs from the space probes *Spirit* and *Opportunity*, a child might ask an elder, "Why is the sky red?"

Sky color can also be altered by clouds, which reflect, scatter, and absorb wavelengths in a variety of ways, as can falling rain and snow. The color of surfaces, natural and human-altered, may tinge the horizon with reflected color. More and more, lights, particularly the yellowish mercury lamps of cities and other large infrastructures, alter sky color. City street and parking lot lighting often casts a colored pall on the underside of low-lying clouds and fogs.

Sky color can reveal important information about the nature of the air above us. Milky blue skies, common during the hazy days of summer, are usually an indicator of high humidity levels or heavy pollution within an air mass. Winter's crisp blue skies indicate that cold, dry arctic air is likely present. As a result, many folk sayings use color as a weather prognosticator, such as "Red sky at night, sailors' delight." This saying is based on the fact that the red sky to the west is usually a cloudless sky, indicative of the passage of a weather system that had brought rain and perhaps stormy conditions, but now brings clearer weather.

Twenty-four

The Morning Dew

As the sun rose across the landscape, its first rays reflected off the ground, giving a bejewelled light. Automobiles were beaded with moisture, despite the fact no rain had fallen overnight. As the saying goes, dew had fallen overnight, though as we shall see, dew does not actually *fall*; it just appears when conditions are right, like a fairy-book creature.

With my training as a weather observer, seeing morning dew tells me several things about the previous night. Most important, it signifies that the near-surface temperature had fallen during the night to below the *dew point*, that temperature at which saturation would occur. Second, the probable cause of that temperature drop was a clear nocturnal sky that allowed the surface heat to escape and greatly cool the surface air temperature. Most likely, winds were calm or light as well, and the air mass may have had low humidity content away from the surface.

Humidity: Is It All Relative?

As I explain the formation of dew, this is a good place to tackle some misunderstandings over a commonly used weather term—*humidity*—which many use indiscriminately when they talk about the amount of water vapor in the air.

Let's start the discussion with water. Under normal Earth conditions, water may exist in one of three states: solid, liquid, or gas. Solid water is ice, liquid water we call water, and gaseous water we call water vapor, which, when found in the gaseous mixture we call air, is also termed humidity.

Humidity, according to meteorologists, can be *absolute* or *relative*, giving us the terms *absolute humidity* and *relative humidity*. In our fast-paced media world, the adjective preceding *humidity* is often dropped, leaving us to figure out which "humidity" is being talked about. Weathercasters, including those formally trained in meteorology, are often the worst offenders (mea culpa, as well) because other professionals usually understand which is meant by the context.

Absolute humidity is the mass of water vapor contained in a volume of air, that is, the *density* of water vapor, generally expressed in grams per cubic metre. It is like saying there are X grams of cashews in a jar of mixed nuts (too darn few). Absolute humidity is commonly used by meteorologists because it varies only moderately through an air mass, being added mainly by evaporation, subtracted by condensation, or altered by mixing along the air mass edges.

Relative humidity is the dimensionless ratio, expressed as a percentage, between water's vapor pressure in the air and the saturation vapor pressure for the air temperature. Under normal surface conditions, it is also the ratio of the absolute humidity and the absolute humidity at saturation for the ambient temperature.

The key term in the definition of relative humidity is *saturation*. Saturation is the condition when the partial pressure of water vapor in the atmosphere is at its maximum level for the existing ambient temperature and pressure. At the saturation vapor pressure, equilibrium exists between water vapor and liquid water, and there would be no net evaporation or condensation from a standing water surface. Given the temperature of a volume of air and its pressure, we can easily determine the saturation value. You will often hear saturation called the condition where air contains "all the water vapor it can hold." This is not technically correct, but old metaphors die hard.

Relative humidity is usually what the media mean when they say "humidity." Relative humidity has uses in human comfort determination and many areas of biometeorology, but it can be a confusing measure because its value varies through the day with the air temperature. For example, the morning may have a relative humidity of 78 percent that, as the air temperature rises, drops to 53 percent by afternoon. The absolute humidity, however, may remain unchanged through that day.

Similarly in winter, the outdoor relative humidity may be 63 percent, but when outdoor air permeates our warm homes and offices, the relative humidity level may drop to 35 percent or lower. In this example, the absolute humidity is quite low in the outdoor air, but the saturation value for cold air is also low, thus the outdoor air is more humid, relatively speaking, though the absolute humidity remains essentially the same.

The *saturation temperature* of the ambient air is commonly called the *dew point temperature*, or simply the dew point. Its value is solely dependent on the absolute humidity of the air. You might hear a weathercaster or meteorologist discuss the dew point of a particular air mass. Since dew point, like absolute humidity, varies little within an air mass, it is often a good indicator of air mass type. Dew point temperature is a regularly reported element in standard weather observations, while absolute humidity is not.

A parcel of air can become saturated by increasing its water vapor content through evaporation, or through it mixing with another parcel of more humid air, or by cooling the parcel to its saturation temperature. These processes are at work continually in the atmosphere, but the latter is more familiar to us. It is the usual process for forming dew and often the cause of fog.

Forming Dew

When the ambient temperature of some surface element, such as vegetation or rooftops or car exteriors, falls to the dew point, the surrounding air will become saturated. But when the temperature drops below the dew point, the equilibrium is tilted toward the liquid state, as more water vapor is pressured into the liquid form (condensation) than escapes the liquid state into the vapor (evaporation) and water collects on the surface.

Due to the forces of surface tension, it is easier to move water molecules in and out of the liquid phase when the surface is flatter. However, chemical and physical properties of a wet surface, often aided by gravity, may "pull" the liquid water into rounded shapes that form the dewdrop, enhancing their visibility and longevity.

As I look out across the ground before me, I can see with closer inspection the dew droplets that had glistened in the post-dawn sun like small gems. Many formed on the minute "hairs" covering leaves of grass and bushes. Such hairs not only shape the water into distinct droplets, but also prevent them from rolling or dripping off the leaf surface. The bejewelled beauty of the scene, like many such weather scenes, was of short life, however. Soon, the sun's rays would strike the dewed surfaces, and with its heat would again tip the evaporation-deposition balance, transforming the liquid water back into the vapor state.

Twenty-five

Enter Solar Summer: Beltane

With the first of May, we arrive at the second of the year's cross-quarter days. This one signifies the commencement of solar summer, that period of ninety-odd days when the sun's strength in the northern hemisphere is at its peak and daylight is longest. Solar summer may not be the hottest quarter of the year, but that is due to the earth, lakes, and oceans requiring an added period of time to fully shake off winter's cold—about one month for land and up to two for large water bodies, coastal areas, and high elevations.

When I was growing up, the date was still referred to as May Day, but its significance in the United States waned each year of the Cold War because the Communists had chosen this date to celebrate their ideology with military parades. That May Day lost favor in North America because of this was sad—it had much greater and longer significance as a day for rejoicing life and fertility, one that likely began in prehistory.

Old cultures called the evening preceding May Day New Summer's Eve and *Walpurgisnacht*—a night for fairies, ghosts, and witches to emerge (a spring counterpart to Halloween). They then celebrated *Roodmas*, *Floralia*, and Beltane to honor the regeneration of life that came with the increased sunlight of this peak solar period. In fact, the name Beltane derives from the Celtic word for "brilliant fire." Not only did the date begin the highest sun period, it also signified that killing frosts were likely over, and it was safe to plant certain crops and vegetable gardens. Spring flowers were in bloom, and many animals were in the height of the reproductive processes. For people, the longer daylight hours allowed them to take advantage of the natural light for both work and play.

We have mostly forgotten these festivals in North America, and it's a pity. The day holds much to be thankful for—the photosynthetic factories are in full production and all animal life takes advantage of this abundance.

Lake Breezes

I was born in Chicago within 20 kilometres (13 mi.) of Lake Michigan and, until my relocation to Vancouver Island in 1992, had lived forty-five years within 80 kilometres (50 mi.) of at least one of the Great Lakes. For most of those years, I was within a few hours' drive of two or three of the lakes. Only the shining, big sea waters of *Gitche Gumee*—Lake Superior—have never been within easy reach of my home. Lakes Michigan, Huron (with its Siamese twin, Georgian Bay, which some consider the sixth Great Lake), Erie, and Ontario have all had direct influences on the weather and climate around me through most of my life.

Each season brings slightly different lake influences on weather patterns within the Great Lakes basin. The basin lies under one of the major North American storm tracks, and the interaction between weather systems and the lakes' waters add a further dimension to life here. I believe the richest variety of weather found anywhere on Earth occurs in this region.

In autumn, the lakes moderate the cold air that sweeps south from the Arctic zone, extending the frost-free period and allowing many areas south of the lakes to be prime fruit-growing sectors. The lakes often give birth or new life to storm systems during this season, and a few such storms have blown with hurricane-force winds. Late in the season, outbreaks of cold arctic air generate bands of clouds that sweep onto the shoreline, pelting inland regions with cold rain showers, ice and snow pellets, and finally snow showers and snow squalls. Winter brings an increased frequency of snow squalls as lake-effect snows develop within the howling cold air streams crossing the relatively warm lake waters. The moderation of the cold air mass often spares the lee shores from extreme cold temperatures.

Lake influences during spring and summer can often knock much of the heat and humidity from tropical air moving north from the Gulf of Mexico as it passes over the relatively colder waters. The Great Lakes become popular recreation areas as the cool and drier lake air moves inland on lake breezes, bringing refreshing relief from summer's heat to the millions residing in the region.

For a number of years, the influence of the lake breezes, which form along the Great Lakes' shorelines during spring and summer, were focuses of my professional attention. Lake breezes and their associated circulation systems can have a profound negative effect on air quality. During my years as a meteorologist with Environment Ontario, the lake breeze regime would often cause air pollution alerts in several Ontario locations by trapping industrial emissions within the stable lake air.

The Lake Breeze and Lake Breeze Front

For many, *lake breeze* means a wind flowing off the lake onto the shore. Most meteorologists, however, use the term to define a special, local airflow situation known as the *lake breeze circulation* or *lake breeze regime*. The key conditions for the development of a lake breeze circulation are light regional winds and a strong temperature contrast between air over the cold lake and that over the warm land. These conditions generally occur on spring and summer days when the sun shines strongly and winds from larger-scale weather systems blow lightly. (Such a flow regime is considered a *mesoscale flow*, or perhaps a *microscale flow* on a smaller lake.)

When the air is warmer over the land than the surrounding lake, an area of slightly lower air pressure develops relative to the air pressure over the cold lake. Since wind flows from high to low pressure, a breeze will blow inland across the shore. Typically, the lake breeze will commence in late morning, after the sun has had sufficient time to warm the land, but the timing can vary from day to day depending on the initial land and lake air temperatures and regional cloud cover and wind speed. Late May and early June are prime months for strong lake breezes because the lake/land temperature contrast is often greatest during these months. As the lake waters warm, the lake/land temperature contrast lessens, and lake breezes become less common in late August and September.

When the lake breeze moves inland, it pushes the colder lake air in like a cold front. This *lake breeze front* is narrow in width and can be slow moving, particularly later in the day, when it approaches its greatest inland penetration from the shore. It is often possible to walk across the frontal boundary from the land air mass into the lake air mass and feel the differences between them. On the land side of the front, the air is warm, and winds are generally light and usually variable. On the other side, the air is distinctly cooler, with wind blowing off the water, often being much gustier. Later in the day, the differences may become less apparent, with the wind shifting to parallel the coastline rather than coming in perpendicular to it. This is due to the influence of the Coriolis effect. Thus in late afternoon, with the lake to your right, the lake breeze should be at your back. The lake breeze front may also be delineated by a line of cumulus clouds along it, which can grow into thunderstorms under optimal conditions.

I had the opportunity to experience the contrasts on opposite sides of the lake breeze front during a field study in 1970, while I was a graduate student at the University of Michigan. We were investigating the lake breeze character and its distance of penetration along the western shores of Lake Erie. We had positioned our monitoring equipment such that the front would travel inland past several stations but infrequently go past all of them. Thus, as I moved inland from the lakeshore, servicing the instruments, I was able to feel the difference between the two air masses. On several occasions, I was present at a station when the front passed by and could feel the changing conditions while seeing its distinct signature being written on the instrument charts.

Later in the day, when the temperature situation reverses and cold air is found over land and relatively warm air over water, a reverse flow known as the *land breeze* may occur, blowing off the land and out over the lake.

Lake Breeze Thunderstorms

The lake breeze front can also trigger thunderstorms where the cold lake air pushes the warmer land air upward along the frontal boundary. Such storms can become severe, particularly when separate lake breeze fronts approach from opposite directions and produce a strong convergence zone between them. Converging air can only go upward where they meet, and such strong upward motions are very conducive for thundershower and thunderstorm formation.

The region of southwestern Ontario along a line from Detroit to Toronto is bounded by Lake Huron to the north and Lake Erie to the south. Here, lake breezes frequently move off each lake during hot weather, and the breeze fronts converge somewhere over the region. Canadian meteorologists during the late 1990s undertook a field study dubbed Project ELBOW (Effects of Lake Breezes on Weather), whose preliminary results suggest that thunderstorm formation and severity associated with lake breeze fronts may be more significant than previously believed. As a result, local forecasters began to pay more attention during the warm season for the development of such conditions.

One severe storm event documented by the researchers in July 1997 was triggered by lake breezes flowing off Lakes Erie and Huron. They closely observed a thunderstorm cell develop between the converging lake breeze fronts and grow into a severe storm in about twenty minutes. A spotter team chasing the storm experienced very heavy rainfall and observed downed trees and flooded roads and fields. From local radar measurements, they estimated that 200 millimetres (8 in.) of rain fell at the small Ontario community of Punkeydoodle Corners over a five-hour period due to this stationary storm cell formed between the lake breeze fronts.

While lake breezes and attendant weather events are well observed and documented around the Great Lakes, all large, and many smaller, lakes also develop these regimes, including Lake Winnipeg, Lake of the Woods, Great Salt Lake, Lake Champlain, and the Finger Lakes of New York State.

Twenty-seven

Wind from Sea, Wind from Land

Winds laden with moisture and rain blow in from the sea across the east Asian land mass in summer, while, in winter, dry winds from the interior move out toward the sea. These winds are known locally as *monsoons*, meaning "seasons." Just as there are annual wind direction reversals on a continental scale, there are also daily wind reversals across the land-sea boundary on a regional scale, similar to the lake breeze circulation regime I described earlier. The wind that moves inland daily with cool, moist air is called the *sea breeze*, and that which moves seaward is the *land breeze*.

The ancient Greeks were the first to write extensively of the sea–land breeze rhythm. Homer, in the *Odyssey*, related that both Odysseus and Telemachus set sail after dark to take advantage of the land breeze blowing out to sea. Plutarch spoke of the Athenian commander Themistocles using the onset of the sea breeze, which produced rough seas in the Bay of Salamis, to defeat the Persian fleet. Persian ships could not be maneuvered in the rough seas as well as the smaller vessels used by the Greeks, giving the Greeks the decisive tactical advantage. Aristotle in *Problemata* and Theophratus in *On the Winds* attempted to describe the genesis and nature of the land and sea breezes. They both considered the land breeze to be the dominant partner and the sea breeze only the reflection of the land breeze off obstacles such as islands and coastal hills. They believed that the alternating current, as they called the sea breeze, could not blow across open sea, where no obstacles from which to rebound existed.

A True Child of the Sun

On the rugged Greek coast, such conclusions as to the relative strength of the land and sea breezes are quite justifiable due to the enhancement of the land breeze and weakening of the sea breeze by the land's seaward slope. In general, however, the sea breeze is the stronger of the two winds, especially along those tropical coasts flanked by cold ocean currents, for the sea breeze is a true child of the sun. The genesis of the sea-land flow pattern depends upon the formation of a pressure gradient across the land-sea boundary, with the higher pressure located over the sea. This gradient of pressure is greatly dependent not only on the temperature difference between the land and sea surfaces, but is also influenced by the strength and direction of the large-scale wind patterns, roughness of the terrain, curvature of the coast, and moisture conditions over the land.

Ideally, the picture develops this way. As the day dawns, coastal skies are cloudless or nearly cloudless, and the wind induced by large-scale weather patterns is light. As the sun rises, the increased solar energy flux heats Earth's surface, which, in turn, heats the atmosphere's lowest layers. At sea, however, the radiant energy received is rapidly dispersed by a combination of turbulent mixing due to winds, waves, currents, and the capacity of the water to absorb great quantities of heat with only slight temperature alteration. Thus, the air over land warms faster than that over the sea surface. Since warmer air is lighter air, the pressure over land becomes less than that over water, the average value of this difference being, during the sea breeze regime, about 0.1 kilopascals (1 mb).

A few hours after sunrise, the pressure gradient will have built up sufficiently to allow the sea breeze to begin moving inland. As this happens, the cooler sea air advances like a cold front, characterized by a sudden wind shift, a drop in temperature, and a rise in relative humidity. A temperature drop of 2 to 10 C° (3.6 to 18 F°) within fifteen to thirty minutes is not an uncommon occurrence as the sea

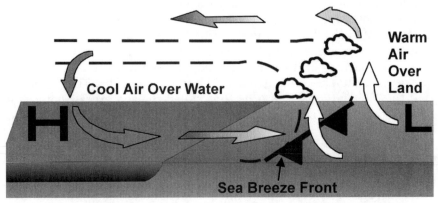

Sea breeze circulation develops when higher pressure over cool water moves inland.

breeze front advances. Thus, in the tropics, the sea breezes make coastal areas more comfortable and healthy for human habitation than the inland regions. For this great service, English colonists overcome by the tropical heat in coastal Africa have bestowed on the sea breeze a special name—the Doctor—and welcomed its coming.

From the time of the sea breeze front passage until late afternoon, the wind will blow inland at speeds of 13 to 19 km/h (8–12 mph), and occasionally as strong as 40 km/h (25 mph). At first, the wind blows perpendicular to the shore, but as the day wears on, friction and Coriolis effects act to veer the wind until it parallels the coastline. The landward penetration of the sea breeze reaches 15 to 50 kilometres (9–30 mi.) in the temperate zones and 50 to 65 kilometres (30–40 mi.) in the tropics. By late afternoon, the strength of the sea breeze slowly diminishes as the influx of solar energy lessens. The decay of the circulation pattern occurs first at the shoreline and proceeds farther inland.

The Land Breeze

As the sun sets, cooling begins along the surface of the land and sea. Like daytime heating, cooling occurs at different rates over water and land. The rapidly cooling land soon has a higher air pressure over it relative to that over the sea, and the air begins to flow seaward, down the pressure gradient. This is the land breeze. It too is influenced by the roughness of the coastline, the strength of the large-scale winds, and coastal configuration. Unlike the sea breeze, the land breeze is often weaker in velocity and less commonly observed, but its flow can be greatly enhanced by coastal topography, particularly off coastal mountains. In such terrain, the descending colder air is accelerated by the force of gravity as it moves downslope. The land breeze is often dominant for only a few hours, and its direction is more variable than

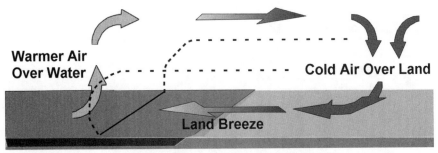

Land breeze circulation forms when cold air over land flows offshore toward lower pressure over warmer waters.

the sea breeze. Nevertheless, the land breeze can penetrate the marine atmosphere for 10 kilometres (6 mi.) seaward.

Climatology of the Sea and Land Breezes

The sea breeze is most common along tropical coasts, being felt on about three out of every four days. The warmer temperatures, increased solar radiation, and generally weaker prevailing winds in the low latitudes promote the development of the sea breeze. In general, the climatic significance of the sea breeze decreases with latitude. In temperate regions, it is usually a phenomenon of late spring and summer, when atmospheric conditions (higher land temperatures, cooler water temperatures, and weaker large-scale winds) are most favorable to the formation of the thermally induced sea-land circulation system. Along coasts with steep shorelines or volcanic island coasts, however, the land breeze may be the dominant partner, with speeds in excess of 32 km/h (20 mph). The land breeze may also occur in the temperate regions during the cold season, especially where a warm current flows along the coast.

Twenty-eight

A Cloud-Watching Kind of Day

This morning, the sky dawned predominantly clear, but by late morning, the first puffy heaps of cumulus began to grow over the Olympic Mountains. Their presence drew my gaze as they slowly migrated northeast. I love cloud watching and spend as much time doing so as I can. (If bird watchers like being called birders, should we cloud watchers be called clouders?) May is a great month for this pastime. By then, the major storm tracks have moved northward, and large, beclouded storm systems are infrequent visitors. As a result, low, expansive stratiform clouds relinquish the airy stage to the cirrus and cumulus families. The warm air and warmer sun provide excellent breeding grounds for towering cumulus

Wave clouds extending from southwest to northeast over Washington, District of Columbia.
(NATIONAL OCEANIC AND ATMOSPHERIC ADMINISTRATION/U.S. DEPARTMENT OF COMMERCE,
NOAA HISTORIC NWS COLLECTION)

119

growth. Here on the Pacific coast, the marine air shoved up over the mountains also supplies an endless parade of cirroform clouds whose ice crystals often surprise us with small gems of halo, iridescence, or parhelia (see the chapter on Halos and Sundogs, page 193).

Clouds provide such infinite variety of scenes as they interact with sunlight and the influences of dust and water, wind, and terrain. Cloudscapes fuel our imagination, producing senses of wonder and beauty, and, at times, dread. No wonder clouds have been an inspiration to all forms of the arts. They have been celebrated in song and verse. The great composers have composed sound poems about clouds, while the great masters have painted the many moods of clouds on canvas. From Claude Monet to Percy Bysshe Shelley, from Henry David Thoreau to Ludvig van Beethoven to Claude Debussy—all have celebrated clouds.

Naming Clouds

Before the nineteenth century, most weather observers believed that clouds were too transient, too changeable, and too short-lived to be classified or even analyzed. With few exceptions, no cloud types were even named; they were just described by their color and form as each individual saw them: dark, white, gray, black, mare's tails, mackerel skies, woolly fleece, towers and castles, rocks, and ox-eyes. Clouds were used in a few instances as forecast tools in weather proverbs, but mostly with regards to their shape or state of darkness or color:

"Red cloud at dawning, shepherd take warning."

This cumulonimbus cloud is about to anvil into a thunderhead as ice crystals form in upper levels. (KEITH C. HEIDORN)

An elk is silhouetted against an altocumulus cloud deck outside Jasper, Alberta. (KEITH C. HEIDORN)

"Mackerel skies and mare's tails, make lofty ships carry low sails."

Then two cloud classification schemes were independently developed within a year by French naturalist Jean Baptiste Lamarck and English pharmacist Luke Howard. Both schemes were likely inspired by the work of the great Swedish taxonomist Carl von Linne, known today as Linnaeus. Linnaeus's systematic classification scheme for all life forms was one of the most significant scientific milestones of the eighteenth century and was adopted by scientists and naturalists around the world.

Lamarck was the first to present his cloud classification system, publishing a paper titled "On Cloud Forms" in 1802 in the third volume of his *Annuaire Météorologique*. Lamarck realized the importance of clarity in observing meteorological phenomena: "It is not in the least amiss for those who are involved in meteorological research to give some attention to the form of clouds; for, besides the individual and accidental forms of each cloud, it is clear that clouds have certain general forms which are not all dependent on chance but on a state of affairs which it would be useful to recognize and determine."

He initially proposed five main types of clouds "related to general causes which are easily ascertained." The types were hazy (*en forme de voile*), massed (*attroupés*), dappled (*pommelés*), broom-like (*en balayeurs*), and grouped (*groupés*).

Three years later, Lamarck devised a more detailed classification scheme involving twelve forms. His system, however, did not make an impression on scientists and naturalists of the day and does not seem to have been used by anyone except Lamarck himself. One reason for this, put forward in the preface of the *International Cloud Atlas* published by the World Meteorological Organization in 1939, was that his choice of peculiar French names was not readily adopted in other countries.

During the winter of 1802–03, Luke Howard presented a paper to the Askesian Society, of which he was a founding member, titled "On the Modification of Clouds." ("Modification," at that time, was used in the sense of "classification.")

Howard proposed that several simple categories identified the complexity of cloud forms. The great leap he took was to name his descriptive categories in Latin (as Linnaeus had done with the plant and animal kingdoms), the language of scholarship, thus transcending national and language borders. Unlike Lamarck's name choices, these were understandable to all European-derived cultures (and in non-European lands, where the Catholic Church had made inroads, bringing Latin to local scholars). It also helped that the system was both very simple and nearly all-encompassing.

Howard believed all clouds belonged to three distinct groups:

Cumulus (Latin for "heap")	"Convex or conical heaps, increasing upward from a horizontal base—Wool bag clouds."
Stratus (Latin for "layer")	"A widely extended horizontal sheet, increasing from below."
Cirrus (Latin for "curl of hair")	"Parallel, flexuous fibers extensible by increase in any or all directions."

To denote a cloud in the act of condensation into rain, hail, or snow, he added a fourth category:

Nimbus (Latin for "rain")	"A rain cloud—a cloud or systems of clouds from which rain is falling."

According to Howard, "While any of the clouds, except the nimbus, retain their primitive forms, no rain can take place; and it is by observing the changes and transitions of cloud form that weather may be predicted."

Clouds could also alter their forms, thus, Howard reasoned, when cumulus clouds bunched together so that they crowded the sky, they became *cumulo-stratus*. Similarly, he defined other intermediate categories of transformation:

Cirro-cumulus	"Small, well defined, roundish masses increasing from below."
Cirro-stratus	"Horizontal or slightly inclined masses, attenuated towards a part or the whole of their circumference, bent downward or undulated, separate, or in groups, or consisting of small clouds having these characters."

Howard's work made a big impression on those interested in the sky, particularly after his papers were reprinted in Thomas Forster's *Researches about Atmospheric Phaenomena* in 1813. The classification system quickly gained wide acceptance both in Britain and other countries. Among its biggest supporters was German poet, philosopher, and scientist Johann Wolfgang von Goethe. Goethe wrote in a letter to Howard (translation found in Howard's notebook): "how much the Classification of the clouds by Howard has pleased me, how much the disproving of the shapeless, the systematic succession of form of the unlimited, could not but be desired by me, follows from my whole practice in science and art." The detailed description of cloud formations in Howard's work also appears to have influenced many Romantic era painters, notably Joseph M. W. Turner and John Constable of England and Caspar David Friedrich (through Goethe) in Germany. They used Howard's descriptions to depict clouds with greater detail and accuracy. Turner first learned of Howard's work through the second edition of Forster's book in 1821, and it inspired him to paint a series of cloud studies.

Howard's classification was accepted almost intact by the meteorological community in 1874, with a few additional terms, such as *alto* (meaning "middle"), for clouds located at intermediate altitudes. When the International Meteorological Committee adapted and adopted Howard's classification scheme for international weather observations, it commissioned a cloud atlas so observers could visualize the descriptions given in conjunction with corresponding photographs. That atlas was published in 1896.

Since then, a number of official and unofficial cloud atlases have been published, and cloud atlases have been incorporated in many field guides to the atmosphere and weather. I liken these to the bird field guidebooks that allow the user to identify the "species" seen. But, like birds, each cloud displays individual characteristics when viewed over a period of time. If one is well positioned and patient, the full life cycle of individual clouds can be viewed.

A fallstreak of ice crystals produces the "mare's tail" of typical cirrus cloud form.
(KEITH C. HEIDORN)

Twenty-nine

Let's Rumble: Thunderstorms

A hot summer's day had finally ended, only to usher in a hot summer's evening. The house was too hot and stuffy to stay inside, so our family sought the little relief the outdoors could offer. As we sat in the yard, I began to notice the northwestern sky glow with random flashes of light—but no thunder followed. "Might get a thunderstorm tonight," I forecast. "Hope so," my brother replied.

At any given moment, tens of thousands of thunderstorms rumble across the face of Planet Earth, an estimated 16 million each year. While some become violent and throw injurious and damaging temper tantrums, most produce a show that entertains and fascinates all within their sky-vaulted theater.

Some require sunny days to form, others advancing cold air, and a few need mountains, but two ingredients are essential: moist air and a rapidly moving upward airstream. Every thunderstorm contains an updraft at its core, a channel of rapidly rising air moving straight up the storm's center. Other factors within its environment that can enhance the moisture content, or the rate of the updraft, or height of the updraft's reach will give added body to a thunderstorm, even setting off its more violent tendencies.

"Garden-variety" single cell, air mass thunderstorms pop up on many a sunny, warm spring or summer afternoon. They generally light the sky with lightning, rumble with thunder, drop light to moderate showers, and bring gusty winds in their passing.

Their life cycle begins with the overheating of moist air at Earth's surface by strong solar radiation. When surface air is heated, its density decreases, and parcels of warm air begin to rise into the relatively colder air above. As the air rises, it cools and, if initially warm and moist enough, eventually reaches an altitude where condensation of its water vapor begins. This is the *lifting condensation level*, which is locally fairly uniform in height, as you can

The updraft base of well-developed thunderstorm during a Michigan severe-weather event.
(KEITH C. HEIDORN)

see by the uniformity of the bases of cumulus clouds formed during an afternoon.

If the heating is not great enough, the moisture content of the rising air too low, or the atmosphere into which these buoyant parcels rise not conducive to further ascent of the parcels, we get small, fair-weather cumulus (*cumulus humilis*) and nothing larger.

The *cumulus stage* can be either the end of the process or the initial stage for those clouds destined to form thunderstorms. If conditions are just right, the cumulus clouds continue to grow in height as the core updraft rapidly ascends, gaining more heat energy as its water vapor is converted into liquid water. This added heat keeps the updrafts alive, and they ascend until all buoyancy is lost.

When the updrafts in the developing cloud reach their maximum altitude, usually around 12 to 14 kilometres (7–9 mi.), the thunderstorm reaches its *mature stage* of development. The upper reaches of the cumulonimbus cloud take on its characteristic anvil shape, as strong upper-level winds spread ice crystals that have formed in the top portions of the cloud horizontally downwind. At this time, the air in the updraft can no longer rise and changes its direction to become a downdraft. Within the up- and downdrafts, precipitation begins to form as cloud droplets and ice crystals combine through collision and coalescence and, when the resulting raindrops become heavy enough, descend to Earth. A thunderstorm's mature stage is characterized by lightning and thunder, rainfall, and gusty surface winds whose direction is highly controlled by the storm itself.

Having reached the mature stage, the thunderstorm begins to decrease in intensity and, after about half an hour, enters the *dissipation stage*. Air currents within the convective storm that were once updrafts are now mainly downdrafts as the storm's supply of warm, moist air from below is depleted. Within an hour, the storm has run its course, and all precipitation has stopped.

Such air mass storms appear isolated from one another as they form and die, but under certain atmospheric conditions, cells merge and form lines of thunderstorms, often bringing violent weather: hail, strong winds, downbursts, and at times, tornadoes. Air mass storms may appear to be long lasting but are, in fact, a series of individual cells, each forming within the dying thunderstorm's region of influence. Eventually, the solar heating that spawned these thunderstorms wanes, and the atmosphere quiets for the night. If, however, atmospheric conditions are ripe for thunderstorm development and the solar energy was a trigger rather than the full cause, thunderstorms can continue to rumble through the night, producing light shows that entrance even the less-dedicated weather observer.

Cumulonimbus mammata are dramatically lit by the setting sun over the American Midwest. (KEITH C. HEIDORN)

Thirty

Cloudbursts of Many Stripes

Have you ever gone out for a walk on a gorgeous spring or summer day and been caught in a torrential cloudburst? It's bad enough to be caught in the rain, but cloudbursts can leave you drenched to the bone—and, in some cases, concerned for your life.

Cloudbursts or *downpours* have no strict meteorological definition. The term usually signifies a sudden, heavy fall of rain over a short period of time. Some observers suggest a rainfall rate in excess of 25 millimetres (1 in.) per hour constitutes a downpour, but when you're drenched, the exact amount often does not matter all that much.

We do know that most cloudbursts come from convective, cumulonimbus clouds that form thunderstorms and that the air is generally rather warm in order to contain the amount of moisture needed for a heavy downpour. Besides providing the proper conditions to spawn large quantities of liquid water drops, cumulonimbus clouds have regions of strong updrafts that hold raindrops aloft en masse and can produce the largest raindrops—those greater than 3.5 millimetres (0.14 in.) in diameter. These updrafts are filled with turbulent wind pockets that toss small raindrops around with surprising force. Within the turmoil of the randomly moving drops, there are more collisions among the drops than on a bumper car ride, and many of those close encounters result in their conglomeration into new, larger drops.

Eventually all updrafts collapse, and when they do, the upheld raindrops descend unimpeded toward the surface, often forming a strong downdraft, such as a *downburst* or *microburst*, in the process, an event that appears as if the cloud has burst open like a soggy paper bag. So, not only are the larger drops falling with a terminal velocity of around 12 km/h (7 mph), they have the added giddy-up of the downdraft speed, which can easily exceed 80 km/h (50 mph).

The resulting rainfall is a torrent of water, large raindrops falling at high speed over a small area. The force and quantity of such downpours can be damaging to vegetation, small animals, and property. When the speed of water accumulation on the ground exceeds the surface's ability to absorb it, localized flooding will occur in low-lying terrain. In hilly or mountainous terrain, the runoff of water can congregate in stream beds or canyons and cause deadly and damaging flash flooding.

Here are some cloudburst milestones:

World-record Cloudbursts

Duration	Accumulated Depth	Location	Date
1 minute	38.1 mm (1.5 in.)	Barot, Guadeloupe	26 November 1970
5 minutes	61.72 mm (2.4 in.)	Port Bells, Panama	29 November 1911
15 minutes	198.12 mm (7.8 in.)	Plumb Point, Jamaica	12 May 1916
20 minutes	205.74 mm (8.1 in.)	Curtea-de-Arges, Rumania	7 July 1947
40 minutes	234.95 mm (9.25 in.)	Guinea, Virginia, US	24 August 1906

Huge cumulonimbus clouds that tower miles into the sky not only bring cloudbursts of rain on occasion, they may also bring downbursts of wind and even heat. You may have heard of downbursts, or their little brother, the microburst. These strong, damaging winds drop from severe thunderstorms, and when they impact on the ground often give the impression a tornado has struck. *Heatbursts*, on the other hand, are rarer, and in all probability, these sudden rises in temperature will make the news only as a weather oddity. Both, however, have similar origins—downward-moving air from a thunderstorm's core—and affect a rather small area.

Downbursts

The late University of Chicago storm researcher Dr. Ted Fujita, who developed the Fujita Tornado Damage Scale or F-Scale, advanced our knowledge of tornadoes so significantly that he was known by his colleagues and the public as Mr. Tornado. Fujita was more than just a tornado expert, however, but also a pioneer in research into all damaging thunderstorm wind events. It was he who first coined the term "downburst."

Fujita first became aware of these winds while investigating the crash of Eastern Airlines Flight 66 at John F. Kennedy Airport in New York on 24 June 1975. After reviewing the weather data at the time of the crash, he believed severe downdrafts were responsible for the accident and needed a simple term to describe the phenomenon. A downburst, Fujita explained, was a straight-direction surface wind in excess of 62 km/h (39 mph) caused by a small-scale, strong downdraft from the base of convective thundershowers and thunderstorms. Later investigations into these thunderstorm phenomena led him to define two sub-categories of downbursts in 1981: the *macroburst* and microburst.

Macrobursts are downbursts with winds up to 188 km/h (117 mph) that spread across a path greater than 4 kilometres (2.5 mi.) wide at the surface and last from five to thirty minutes. Microbursts, on the other hand, are confined to a much smaller area, less than 4 kilometres (2.5 mi.) in diameter from the initial point of downdraft impact. An intense microburst can push damaging winds near 270 km/h (170 mph) and often lasts for less than five minutes.

Downbursts of all sizes descend from the upper regions of severe thunderstorms when the air accelerates downward through either exceptionally strong evaporative cooling (*dry downburst*) or by very heavy rain that drags dry air down with it (*wet downburst*). When the rapidly descending air strikes the ground, it spreads outward in all directions, like a fast-running faucet stream hitting the sink bottom.

When a microburst wind hits the ground, it can flatten buildings and strip limbs and branches from trees. After striking the ground, the powerful outward-running gust can wreak additional havoc along its path. Damage associated with a microburst is often mistaken for the work of a tornado, particularly directly under the microburst. However, damage patterns away from the impact area are characteristic of straight-line winds rather than the twisted pattern of tornado wind damage.

Microbursts are a particular threat to aircraft landing or taking off because they can force the aircraft rapidly toward the ground or push either the aircraft's tail or nose down, causing the pilot to lose control. In the past, microbursts were generally undetectable because of their small size. Today, however, Doppler radar allows for fine-scale coverage of wind fields around airports. Microbursts can also overturn boats on lakes or quickly fan a forest fire into a raging inferno.

Heatbursts

Another "burst" that may drop from a thunderstorm is the heatburst. If you have experienced thunderstorms, you are likely quite familiar with the cold wind gusts flowing from the thundercloud that frequently bring welcome relief from oppressive summer heat. Most thunderstorms produce these cool gusts of wind, but occasionally, a very hot blast of air descends instead, a heatburst. Meteorologists are still unclear on the exact mechanism behind heatbursts; however, storm researchers believe they originate in the highest levels of tall thunderstorm clouds and are similar in nature to the dry microburst.

Here is the scenario envisioned for the majority of heatbursts. As the sun sinks in the evening sky, its heating of the ground surface first diminishes, then ceases altogether. Without solar heating to lift them, the warm, moist, rising columns of air that fuel the thunderstorm cell terminate, shutting off the storm's main core updraft. Without the updraft, raindrops in the storm's upper levels can no longer be held aloft, and the storm begins to collapse. The raindrops fall into the cool, dry air entering the storm from its rear, where they evaporate, further cooling the air.

This mass of cold air is now very heavy relative to the air around it and plunges rapidly earthward. Usually, this sequence of events produces an outburst of cold air from the storm. However, should the plunging air drop from a very high altitude, say over 6,100 metres (20,000 ft.), it warms significantly by compression (like the air inside a bicycle pump) during the descent. Such compressional warming makes the air lighter—more buoyant—which should arrest its descent, but in this instance, the downdraft's momentum is too

An approaching thunderstorm with a lead gust front forms a flat "shelf cloud" over Brookhaven, New Mexico. (NOAA PHOTO LIBRARY, NOAA CENTRAL LIBRARY, OAR/ERL/NATIONAL SEVERE STORMS LABORATORY)

great to be halted, and the air slams into Earth's surface, spreading outward as a hot, dry gust of wind.

Any tall, dying storm can produce a heatburst if sufficient evaporation takes place high in the cloud, so heatbursts can also form late at night and early in the morning. A heatburst's intensity depends on the collapsing storm's initial size and the degree of warming before the air hits the ground.

Heatbursts are considered unusual events, but no one knows how frequently they actually occur. Many likely strike in the vast, open lands of the mid-continent, from Texas to the Canadian prairies, but are never detected due to the scattered population and lack of weather instruments in the region to observe them. When they do occur, the surface temperature rise can be dramatic. One heatburst, caught on the Oklahoma Mesonet weather-observing network on 22 May 1996, spiked the temperature from 31 to 39°C (88 to 102°F) over twenty-five minutes. As the temperature rose, wind gusts reached 170 km/h (105 mph) within the ten-county coverage area. Several other, smaller heatbursts have been caught by the Mesonet over the past three years.

Thirty-one

Winds of the Day

On this early June morning, nary a breath of air stirs the now-verdant trees surrounding me. The only motion arises from the busy birds dashing from branch to branch searching for food to feed their chicks. June is a month in which warm high-pressure cells, both of continental and maritime origin, begin to dominate weather patterns over most of North America. Those frontal wave storms associated with the polar front have retreated far northward as the temperature differential between polar and tropical latitudes lessens with the lengthening days of northern summer. Weather map lows are now mostly shallow valleys between the sluggish high-pressure cells sprawling over the map or regional dimples of thermal origin formed over the western American desert.

I aim my weather eyes this day a little closer to home, focusing into the realm of *micrometeorology*—that field of study that looks into those atmospheric processes whose extent is less than a few kilometres in all directions.

When this morning dawned, the skies were mainly clear; I know that from the telltale patches of dew I observed upon awakening just after the sun had cleared the treeline. Soon, its warming rays strike the ground at a more concentrated angle and daytime heating will begin.

Bubbles, Thermals, and Turbulence

By mid-morning, pavement, east-facing rock outcrops, and patches of bare soil have become significantly warmer than the surrounding meadows, forests, and lake. The air next to these hot surfaces also gains heat and in the process becomes less dense, a physical process known as *Charles' Law*: as temperature increases, density decreases at constant pressure. Eventually, such heating produces nearly invisible "bubbles" of light, warm air and later plumes of hot air that rise skyward. I say "nearly invisible" because the observant eye can catch them altering any light rays that pass through, producing a shimmering or blurring of images behind them.

Likely, I will also see small fair weather cumulus clouds pop up where the thermal plumes reach their condensation level. If conditions of heat and humidity are conducive throughout the atmosphere, I may even be treated to a thundershower late in the day. But for now my interest is riveted within the surface layer of the atmosphere—that within a building's height of the ground.

With the heating of the air adjacent to Earth's surface, the lower atmosphere becomes *unstable*, a term used to describe the situation where warm, light air is located beneath layers of relatively cold and dense air. The condition is unstable because the lighter air does not long remain in place, but is soon forced rapidly upward by the forces of buoyancy. Once displaced from its original position, warm air keeps ascending until it has lost all its buoyancy, its place taken by cooler air descending into the "void" and completing an overturning of the local atmosphere.

As these bubbles rise, the local air pressure of their spawning ground drops slightly relative to the air around it. As a result, a small pressure gradient is established between the thermal low-pressure spot and the surrounding air, and, as we know, a pressure gradient is the driving force behind the wind. Thus, air from the cooler surroundings blows into the low-pressure spot, where it is eventually heated until it rises. Between the rising hot bubbles, cooler air sinks to replace the rising air, bringing with it cooler temperatures than the ascending bubbles.

If we look out over a sunlit landscape and use a little imagination, we can almost see these bubbles and thermals rising into the sky like chimney plumes on a cold winter's morning: their size and shape varying with the terrain, and their angle of rise varying with the wind speed above the surface. If we could color the cold air blue and the warm air red, we would likely see a number of narrow columns of red air rising swiftly off the surface into the atmosphere and broad regions of blue air slowly sinking in response.

On a larger scale, such conditions can form the lake/sea breeze, upslope breeze, or up-valley wind circulations. On our microscale, rising bubbles and sinking air combined with mixing along the boundary (or *microfront*) between the warm and cold air make the movement of air chaotic, a flow condition called *turbulent*. The local wind thus gusts to higher speeds, then drops—frequently to short-lived lulls—and changes directions incessantly, often with abrupt reversals.

Under low turbulence, both wind speed and direction traces on an anemometer chart meander slowly, if at all, across the display or vary only within narrow confines. But increase the turbulence level, say on a sunny day, and both traces swing across the display as radically as the flight of a perturbed fly. The wind direction often fluctuates through the full compass of directions in a matter of seconds.

The diurnal wind variability, manifest by daytime winds stronger and gustier than nighttime winds, has long been recognized by folk weather observers, although not necessarily with an understanding of the science behind it. They expressed their knowledge in the simple rhyming couplet: "Winds of the day do wrestle and fight, / Longer and stronger than those of the night." The "wrestle and fight" nature of daytime winds is the manifestation of the turbulence level observed in the wind as micro-changes in the heating and pressure fields of the lower atmosphere cause air currents to vary in response. Throughout the day, the solar heating process continues, often altered by the movement across the sky of cumulus clouds, born on rising air bubbles, that cast shadows over the landscape. The wrestling winds gust up, change direction, and then lull for a moment or two. The fight proceeds irregularly until the hours prior to sunset.

Eventually, the setting sun's rays are again spread too thinly to heat the ground above the surrounding air temperature, and the rise and fall of air parcels diminishes, then ceases altogether. Unless perturbed by larger-scale weather patterns, the surface wind speed drops, soon becoming part of the general calmness of evening. With continued clear skies and light winds, Earth's surface will lose heat rapidly and continue to cool. This forms a temperature inversion, cold air underlying warmer air, which further dampens the turbulence in the breeze until it is finally becalmed.

The sun has set, and the day I had awakened to and watched so intently for hours has now ended. Hardly a leaf now rustles in the trees around me, and even the avian activity has ceased for the night. The calmness of the evening ushers in a time for rest, a time for peace, as the waning June moon rises from the horizon.

Thirty-two

Rainbows and Moonbows

In the Book of Genesis, God sends his judgment upon the peoples of Earth with a flood that lasts forty days and forty nights. When the rains end and flood waters finally recede, God forms a rainbow, telling Noah, "I have set my rainbow in the clouds, and it will be the sign of the covenant between me and the earth. Whenever I bring clouds over the earth and the rainbow appears in the clouds, I will remember my covenant between me and you and all living creatures of every kind." (Genesis 9:13–15)

I must admit, at times here on the British Columbia raincoast, I envy Noah having *only* forty days and nights of rain to deal with. But, within those long, wet periods, we are frequently treated to the sight of a rainbow, often several on the same day.

Two items are needed to form a *rainbow*. The first is rain (are you surprised?) or some other source of liquid water drops, and the second is a strong light source.

Rainbows can appear in either the eastern or western skies, but most often we see one when clearing western skies return sunlight to the day and strike the rain falling from clouds moving eastward. Rainbows in the east frequently presage drier weather to come, thus the old shepherd's weather adage: "Rainbow at night, shepherds delight; / Rainbow at morning, shepherds take warning." has some validity for shepherds, sailors, farmers, or whomever the adage is attributed to. When we see a rainbow in the morning hours, it signifies sunlight striking rain falling to the west. Since most weather systems move from west to east, that rain is likely moving toward us, and it is wise to carry an umbrella or raincoat. In contrast, when the rainbow forms at "night," actually from late afternoon to sunset, the rainstorm is moving away, a delight for those wanting dry weather.

When solar rays shine on falling raindrops, they enter the drop and are bent from their straight-line path, a process known as

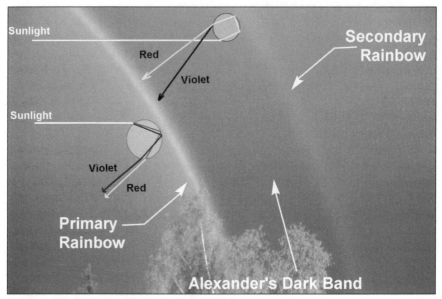

Schematic showing features of a double rainbow.

refraction, which was discussed with regards to mirages, on page 94. If the light rays contain multicolored wavelengths—unadulterated sunlight contains all the colors of the spectrum—they are split by the *refraction* into bands of pure-color wavelengths, much as a prism splits "white" light into the spectrum. These colored rays, traveling in slightly different directions, are then *reflected* off the back surface of the raindrop toward the sun. On leaving the drop, another refraction spreads the colors into a parallel front of beams heading to the observer's eye. Violet light bends the most, emerging at an angle of 40 degrees relative to the incoming sunlight. Red light bends the least, exiting the drop at an angle of 42 degrees. Other rainbow colors emerge from the raindrop at angles between 40 and 42 degrees.

This refraction-reflection-refraction process forms a *primary rainbow*, so called because it is the product of a single reflection of light within the raindrops. Not all the light escapes the drop, however, as some rays are internally reflected off the front of the raindrop for a second internal circuit before leaving the drop to emerge at a different angle (around 51 degrees). This resulting rainbow lies outside the primary bow arch and is called the *secondary rainbow* since it forms from that second reflection. The secondary bow is weaker in brightness (by about 43 percent) than the primary bow and is twice as wide, with its color sequence in reverse order.

The combination of primary and secondary rainbows is called a *double rainbow*. The secondary rainbow lies outside (above) the primary rainbow. In the primary, violet/blue is on the inner edge and red on the outer. In the secondary bow, red is on the inner edge and violet/blue on the outer. Between the two bows, we see a much darker region known as *Alexander's dark band* (for Alexander of Aphrodisias [AD 198–211], a Greek natural scientist who studied rainbows). In this region, most of the reflected/refracted light has been bent along other angles, leaving this region of sky fairly depleted of light rays. In contrast, the rainbow center appears much brighter than the rest of the sky because there, the sunlight reflects directly back from outer and inner drop surfaces.

Seeing Rainbows

Although we usually see a rainbow as a continuous mix of colors, only one color actually reaches our eyes from each drop. We see red light from higher-altitude drops producing the outer ring of the bow, while its violet light rays are directed outside our view. Violet light from lower-altitude drops makes the inner ring visible, while its red light rays are directed below our line of sight. It therefore takes millions of falling raindrops to produce the full color spectrum characteristic of a rainbow arch.

The size of raindrops also influences the rainbow we see. Rainbows with the brightest colors appear when raindrop diameters are between 0.3 and 1.0 millimetres (0.01 and 0.04 in.). If drops are large—1 millimetre (0.04 in.) or more in diameter—red, yellow, and orange colors are bright but blue is weak. As drops get smaller, the red color band weakens in favor of the blues. Very small water droplets—those less than 0.03 millimetres (0.001 in.) in breadth—produce rainbows that are almost white. Such small droplets are found in fog and clouds, thus the bows resulting from them are known as *fogbows* or *cloudbows*.

When sunlight from a sun very near the horizon produces a rainbow, it may not contain the full spectrum of colors when it strikes the raindrops. Often, such light has lost blue and green wavelengths, giving the sun a reddish appearance. Thus, if the light rays entering a raindrop are primarily of one color, say red, they cannot be split further, and the resulting rainbow will be reddish. Similarly, if the sunlight is orange or golden, so too will be any resulting rainbow. I once witnessed a spectacular reddish gold rainbow one December evening, perhaps the single most beautiful skyscape I ever beheld.

Every rainbow forms as part of a circle. How much of the circle we see depends on several factors, including the width of the rain band and the position of the sun relative to our eyes. The center of

Rainbow graces the morning sky over Victoria, British Columbia. (Keith C. Heidorn)

the rainbow circle can always be located by drawing a line from the sun through our eye position to the sky before us. This designates the *antisolar* point. As the solar position rises in the sky, the antisolar point lowers until the rainbow arch sinks beneath the horizon. That is why midday rainbows are uncommon to ground observers, except in high latitudes or during winter. However, a rainbow formed on raindrops falling at lower altitudes can be seen around midday from an aircraft or high mountain peak.

If you and I, standing side by side, notice a rainbow, we are actually seeing two distinct rainbows, one for you and one for me, each produced by a different ensemble of raindrops. In some instances, the drops that produce my rainbow's red band might be producing another color in yours or vice versa—it all depends on positioning.

Other Bow Forms

Since rainbows are composed of light rays, the internal raindrop reflection is not the only possible reflection. Rainbows can be seen reflected in pools of still water, usually small lakes and ponds. The light from a rainbow striking such a reflecting surface produces what appears to be a mirror image of the rainbow arching down to the water surface. But don't be fooled. This *reflected rainbow* is not the reflection of the rainbow we see, but the reflection of an entirely different rainbow formed, again, from another set of light rays and raindrops.

Sunlight reflected off a shiny surface such as a water body that then strikes raindrops can produce a rainbow known as a *reflection rainbow* (a subtly different rainbow than the reflected rainbow mentioned above). Reflection rainbows usually emerge when the sun is low on the horizon and the water surface is very calm, to allow for near-total reflection of the solar light from the water. The raindrops may now be struck by light from two distinct sources: the actual sun above the horizon and the reflected solar orb located below the horizon. As a result, we may see two distinct rainbows, each with a discrete center point. The reflection rainbow always forms higher in the sky than the direct-beam rainbow and has its center above the horizon; consequently, the reflection bow may actually intersect the primary direct rainbow at the horizon.

Another interesting rainbow type is also a reflection rainbow, though we do not usually think of it that way. It is the *moonbow*, formed from the light of the moon, which, of course, is really solar light reflected off the lunar surface. Moonbows are not as spectacular as solar rainbows to our eyes, but to a weather watcher like myself, seeing the rare lunar rainbow would be as thrilling as a birder spotting a rare species. Moonbows form just like rainbows, when moonlight strikes the falling raindrops. Since moonlight is reflected sunlight, it has the same color spectrum and thus forms a multicolored, banded bow. However, a moonbow generally appears rather colorless because human eyes do not perceive colors well at low-light levels. To the naked eye, moonbows appear delicate white, even black, rather than multihued. Observers have described them as eerie and ghostlike and rather faint. A camera, on the other hand, will register all the colors if properly exposed and produce a photograph of a moonbow that looks like a rainbow. Since moonlight is less brilliant than sunlight, about the only time you are likely to see a moonbow is when the moon is full or nearly full. But because they are faint, moonbows are often obscured by air and light pollution present around most cities. For these reasons, moonbow sightings are infrequent.

Another factor working against moonbows is that scattered showers, the best weather situation to see bows, are, in many locales, generally a creature of the day. In mountainous terrain such as the Pacific Northwest, however, showers spawned by moist air moving over mountain ridges are nearly as likely during the day as night. Therefore, I suspect that moonbows may appear more frequently around mountain regions away from the congestion of urban lights, where terrain slopes induce shower formation while leaving clear skies between the showers for moonlight to shine through.

Earlier, I hinted that "rainbows" can form without the presence of rain. Fog and clouds can form bows. Dewdrops are another source of potential rainbow-forming droplets. And likely all of us have seen bows forming from a garden hose or sprinkler spray. But I think the most spectacular "non-rain" bows appear around waterfalls.

My introduction to waterfall bows came at the edge of mighty Niagara Falls. These spectacular bows start, as viewed from the Canadian side, in the gorge, where the drops are below our feet, during the midday and rise in elevation as the sun lowers in the west. The lofted waterfall spray also creates bows in different parts of the falls complex. It is well known that in places such as Kentucky's Cumberland Falls and Africa's Victoria Falls moonbows are frequently seen around the waterfall's mist. Niagara Falls once had frequent moonbows, but they have since been lost, overwhelmed by local urban lights.

SUMMER

DUST DEVILS • SNOW PILLARS • BLIZZARDS • CLIPPERS • SNOWFLAKES • CREPUSCULAR RAYS • SMOG/NOON • HALOS • SUNDOGS • RAINBOWS • LIGHTNING • THUNDERSTORMS

21 June—The Summer Solstice

The date: 21 June. Sometime during this day, the sun will be directly overhead at 23.5° north latitude, the Tropic of Cancer. The date marks the summer solstice, the longest daylight duration of the year and the shortest night over the northern hemisphere. For me, a solar person, this is the highlight of the year, the midsummer's dream of warmth and light. Many recognize this date as the start of the summer season, although cultures more in tune with the sun's celestial position recognize this date as *midsummer*, the central date in that quarter of the year with the most potential sunlight, what I refer to as "solar summer."

On the summer solstice, the sun completes its six-month upward journey through the sky begun on the winter solstice and starts on its return leg downward. For a brief moment, it stops. On this day, all regions north of the Arctic Circle will experience the midnight sun, and for the North Pole, where the "day" lasts one full year, it will be "noon." In other high northern latitudes, the setting sun will hide just below the horizon for a time while bathing the night in neverending twilight. The solar path in these regions resembles a swimmer's shallow dive below the horizon more than a diver's plunge into deep darkness, which is characteristic of equatorial sunsets. The region with *twilight nights* can be found as far as 13 degrees south of the Arctic Circle at midsummer. This region encompasses areas north of 54.5° north latitude, which includes northern Canada, almost all of Alaska, and several major European cities, such as Edinburgh, Stockholm, and Moscow.

Where the sun does set this day, it sets in its extreme northwestward position, having risen that morning as far northeast as possible. You can check this for yourself if you are able to see the point on the horizon where the sun rises or sets each day. Having reached its farthest northeast and northwest locations on the day of the summer solstice, sunrise and sunset locations both move farther and farther south each day thereafter until winter solstice, when the drift stops and the solar trip back north begins. By watching the sunrise and sunset locations change over the next six months or more, you will also notice that the day-to-day changes are greatest around the equinoxes and become very slow around the two solstices—perhaps a better English term might be the "stallstice."

If one does consider the summer solstice as the beginning of the summer season, there is no justification to complain that summers are too short. Since Earth's orbit is an ellipse rather than a perfect circle and orbital speed varies along the path, the astronomical seasons are not exactly one-quarter year (91.31 days) long. Summer, when defined as that astronomical period between the summer solstice and autumnal equinox, is approximately 93 days, 15 hours—the longest of the four seasons in the northern hemisphere.

For climatologists who define seasons based on monthly mean temperatures, the start of summer weather began three weeks earlier, because the three warmest months of the year in many areas are usually June, July, and August. The air temperature lags behind the solar intensity peak because much of spring's solar energy is used to melt snow and ice, heat land surfaces, and warm sea and lake waters from their winter chill. In continental areas such as the Canadian prairie provinces and American plains states, the annual temperature cycle peaks from late July to early August, whereas in the Pacific Northwest, the warmest stretch of days annually occurs from late August to early September. That extra month is required to heat the colder Pacific waters offshore. As a result, for much of the northern Pacific coastline, the three warmest months are July, August, and September.

With more solar energy available and already-warm land surfaces, more precipitation falls across North America during the summer months as showers rather than as steady rain associated with large low-pressure cells. The plant kingdom takes full advantage of this bounty of sunshine and, barring large moisture deficits, pushes the photosynthetic factories into high gear. It is said that during the period surrounding the summer solstice, if you are very quiet, you can hear the corn grow in the fields.

Thirty-four

From Sea Froth to Raindrops

One of my favorite places to contemplate and seek inspiration for my writing is Island View Beach, north of Victoria, which looks east to the continent across the Strait of Georgia, over a garden stone path of islands.

I traveled to the beach one glorious summer day—warm with little wind and a whole menagerie of clouds to stimulate my senses. The tide was just beginning to turn back in, and the smell of the sea was strong. Except for cars entering the park, the only constant sound was that of the local bird population—gulls, crows, Canada geese, and ravens—and the gently breaking surf. The turning tide coupled with waves from a passing fishing boat rolled a small, shushing surf shoreward, breaking in a banner of bubbles on the sand and beach pebbles. Almost immediately, two thoughts popped into my head and they had an interesting connection.

One idea was about the role that breaking waves, and particularly the bubbles formed in the process, play in the formation of clouds and precipitation. The other idea was about books, particularly my favorite weather books, a topic that had been on my mind since reading an article on classic nature writings. The two ideas found their link in one book on my list of weather classics, written by Duncan Blanchard. His 1966 work, *From Raindrops to Volcanoes*, looked at his research into the causes of precipitation, some linked with breaking sea waves.

Blanchard writes that sea salt particles are common natural condensation nuclei for water vapor, the combination of which is responsible for cloud and precipitation formation. Measurements of airborne particles show that sea salt particles can be found as far inland as the midsection of the continental United States. How these particles leave the sea surface to make their way into clouds and rain far inland is the most fascinating part of Blanchard's tale.

As the waves broke on the shoreline before me, I saw bubbles forming, frothing up the surf and giving it that telltale hiss. This is not, however, the only process by which bubbles form before breaking on the water surface. Chemical and biological processes within the water volume can generate gas bubbles that rise to the surface. Another prime mechanism is the impact of precipitation on the sea surface, both raindrops and snowflakes. (We may not think about snowflakes forming bubbles when they strike a water surface, but Alfred Woodcock, who worked with Blanchard at Woods Hole Oceanographic Institute, found them to be very effective bubble producers.) The main producer of sea bubbles, however, is the wind. Whenever the wind speed exceeds 12 km/h (7 mph), the wind interacts with wave crests to form frothy whitecaps. Every patch of frothy sea surface contains thousands to hundreds of thousands of individual bubbles of many sizes, all about to break.

Let's look more closely at this chaotic picture, focusing attention on just one of the many bubbles.

The Life of a Sea Surface Bubble

In whatever manner it is formed, a sea surface bubble traps a very small volume of gas—usually air—surrounded by sea water. Most bubbles are less than 0.5 millimetres (0.02 in.) in diameter. If we assume the bubble formed far enough below the surface, it formed a spherical volume of gas whose shape resulted from the outward pressure of the gas being balanced by the hydrostatic pressure of the surrounding sea water pushing uniformly inward on the bubble. Since the bubble gas is less dense than the surrounding sea water, buoyant forces act to push it toward the surface.

When the bubble reaches the surface, the combination of buoyant forces and momentum pushes against the top surface layer of water molecules that isolate the bubble air from the atmosphere. Eventually, the surface tension of the water surface is no longer sufficient to hold the gas in. At this instant, the bubble bursts. The surface film that had formed the upper bubble boundary explodes upward, spreading like shrapnel into the air. For an instant, the sea surface has a dimple, a depression, where the bubble had been below the sea boundary prior to bursting. With the sudden release of the pressure exerted by the bubble gas on the surrounding sea, the water rushes in to fill the depression. In closing the cavity, a small vertical column of water with a wide base—described by Blanchard as resembling "a miniature Eiffel Tower"—rapidly extends up from the bottom of the collapsing bubble volume. Researchers call this feature the *bubble jet*.

As the bubble jet collapses back into the sea, it ejects up to five

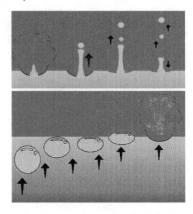

Air bubble rises to sea surface and breaks (bottom panel); the breaking bubble (top panel) forms a rising jet and smaller droplets, which break away from the sea surface.

small droplets that have been pinched off the jet column into the air above. If you have ever felt the tickle on your lips when drinking a fizzy carbonated beverage, you've felt these jet droplets. And you can easily see them by looking closely over the surface of the carbonated drink in a glass.

Many bubble jet droplets (as well as the "shrapnel" of the surface film during the initial bubble break) ejected from the water surface fall back into the sea, forming additional bubbles in the process. Some jet droplets are caught in the turbulent wind field at the surface and are lofted up and away from the waters below. Charles Keith, another member of the Woods Hole research team, determined that bursting bubbles on the sea surface could eject the initial jet droplet as high as 18 centimetres (7 in.) above the surface. Those caught in wind field updrafts may travel great distances and to great altitudes.

Jet Droplets and their Salts

Let's take a closer look at jet droplet composition. Obviously, the jet droplets in this case are composed of sea water, generally from the sea's uppermost layer. Their composition includes water (naturally), a wide variety of sea salts, some organic compounds, and a number of microorganisms—bacteria, viruses, plants, animals—specific to the water body from which the droplet issued. Although the life forms riding the droplet present a fascinating story of their own, I am most interested in the sea salts dissolved in the water.

Sea salts comprise a wide catalog of inorganic and organic compounds, but most of the mass is *sodium chloride* (table salt) and salts formed from sodium, magnesium, calcium, and potassium combined with chloride or sulfate. Those salt particles riding the droplets high into the air can be pushed across the waters as sea spray that hits ships and oil rigs at sea, or inland drifts along the coast forming what coastal ecologists call the *spray zone*. The longer the salty droplets remain aloft, the greater the likelihood that most or all of their water will be evaporated away. When all the water evaporates, minute particles of pure salt are left suspended in the air. Being quite small, they are easily transported hither and yon, up and down, with the vagaries of the turbulent wind. I live about 3 to 5 kilometres (1.8–3 mi.) inland from the nearest shore of Vancouver Island and about 30 metres (98 ft.) above sea level, yet even in the peak of our dry season, my windows become coated with a thin layer of wind-deposited salt. A few kilometres is not a long trip for sea-born salt particles, and some may have escaped from the ocean surface far from the coastline.

Back at the Beach

Back at Island View Beach, as I watched the waves breaking softly over the shore, the winds blew gently from the west and were warm compared to conditions I would find over the open waters before me. The jet droplets formed in the breaking surf's froth were caught in this turbulent wind field where wind, land, and sea meet. Many rode skyward on the warm air. In my mental picture of the process, I saw those minute jet droplets lofted above the cold water surface, headed east toward the mountainous San Juan Islands. Many would become salt aerosols (rather than salty water droplets) by the time they reached these islands, where they were lifted higher aloft as the winds ascended the slopes of the island summits.

Farther east, I saw the winds and the moist, salty air rising even higher as they were forced up and over the continental coastal mountain range. This part of the process I actually did see in the form of cumulus clouds building in the eastern sky. Somewhere over Washington, billions of these salt particles would likely fall to earth, having gathered to form raindrops and snowflakes. But some would escape the deposition and ride air currents ever eastward. Some perhaps may even survive the traverse of the western mountain cordillera to be mixed into air masses descending from polar latitudes and caught in a low-pressure cell to form clouds and precipitation somewhere distant from their departure point. In a few days' time, my Pacific sea salts might fall as rain over my old stomping grounds of Illinois, Michigan, or Ontario. It's nice to remain in contact with them.

And a very lucky few may hitch a ride on the right parcels of air and somehow escape the deposition process for quite a long time, wandering across the globe before finally coming to rest. I like to think a few might even ride full circle and return to Vancouver Island, falling in rain to the beach where I saw them off that day—true world travelers.

Thirty-five

Lightning: A Storm's Flashy Dancer

My most vivid summer weather memories occurred during my mid-teens. One summer in particular shaped the course of my academic and career pursuits for the next four decades. It was the year of my first true love affair—with the sky. That season, some deep curiosity surfaced, urging me to seek an understanding of the weather's ways, an interest sparked by lightning and thunderstorms.

That particular summer in northern Illinois was unusual for its frequency of gentle thunderstorms. The northern Illinois region, my boyhood home, can be the crossroads of some of the wildest thunderstorms, but that summer, they were unusually gentle. Winds were light, at times nonexistent, and rain rarely fell in a downpour. I listened each night for the crackle of static over my transistor radio that indicated the presence of lightning in the area. Comfortably seated at my bedroom window, I gazed into the western sky. Eventually, on the horizon, flashes announced the entrance of the storm line. Then, the low rumble of thunder could be heard. As the storms approached, the lightning changed from sheet lightning, bolts hidden from direct view, to streaks of forked lightning.

I believe lightning bolts are as unique as snowflakes. No two exactly alike. Bolts explode from cloud to ground, to other clouds, or to other portions of the same cloud. What causes this huge spark to form?

Benjamin Franklin and the Early Theorists
Lightning has not yet revealed all its secrets, but we know many of the basic elements of its formation. It results from the attraction between positive and negative charges in the natural environment and the need to alleviate that charge differential.

One of the first scientists to begin to learn lightning's secrets was

American founding father Benjamin Franklin. His legendary kite experiment and invention of the lightning rod are part of American folklore. Franklin's study of the phenomenon in the late eighteenth century was among the first to make the connection between storm lightning and the sparks generated in the early experiments into electrical theory. One of his first papers on the subject, published in 1749, was titled "Observations and Suppositions Towards Forming a New Hypothesis for Explaining the Several Phenomena of Thunder Gusts." In it he states, "Water being electrified, the Vapors arising from it will be equally electrified; and floating in the air, in the form of Clouds or otherwise, will retain that quantity of electrical fire [lightning] 'till they meet with other clouds or bodies not so much electrified; and will communicate . . ."

Franklin devised an experiment whereby he would perch on an electrical stand, a large version of a primitive capacitor called a *Leyden Jar*, from which sparks could be generated and observed while he held a connecting iron rod. Electrical charge from thunderclouds, Franklin hypothesized, would charge the electrical stand. He then hoped to issue an electrical discharge between his other hand and a grounded wire.

Franklin was unable to perform his dangerous experiment himself due to delays in the construction of the electrical stand in Philadelphia. It was, however, successfully performed by France's Thomas Francois D'Alibard in May 1752. In July 1753, Swedish physicist G. W. Richmann, working in Russia, proved thunderclouds contain electrical charge the hard way—he was instantly killed when lightning struck him on his electrical stand.

Before Franklin could attempt his experiment, he thought of another method for proving his hypothesis. He would use a kite. The kite took the place of the electrical stand and iron rod. Legend states Franklin flew his kite during a Pennsylvania thunderstorm in 1752, though many believe it never happened. (Do not try this extremely dangerous experiment yourself, in any event.) As the story goes, history's most famous kite drew sparks jumping from a key tied to the bottom of the kite string to Franklin's knuckles.

Franklin used his knowledge to develop the lightning rod, which sold well across America, and deduced that the lower regions of a cloud were negatively charged. Otherwise, little significant progress in the cause and nature of lightning was made until the late 1800s. The electrical current of lightning was first measured by the German scientist Pockels at the turn of the twentieth century by analyzing the magnetic field induced by lightning currents. Modern lightning research began with the work of Nobel laureate C. T. R. Wilson, who used electric field measurements to study the structure of thunderstorm charges involved in lightning discharges.

Current Theory

The basic process by which lightning occurs is well understood. Regions of positive and negative electrical charge build up and remain isolated from one another by the strong insulating property of air. The separation of charges builds until the electrical potential—the electrical pressure to move electrons from negative to positive—overcomes the resistance of the air to electrical current flow. A simple analogy is the electrical potential, or voltage, between two poles of a battery (one positive, the other negative). No current flows between the battery poles until a conductor connects them. In the case of this simple circuit, electrons will flow as long as the poles are connected by the conducting wire. Since the wire promotes the conduction, the potential or voltage difference need not be great. Most common batteries are less than 12 volts because wire is an excellent conductor.

To overcome the electrical insulating properties of the atmosphere between the two "poles" of charge within a cloud, between clouds, or between a cloud and the ground, the potential must reach millions of volts. Eventually, the air's resistance breaks down under such high voltages, and an ionized path is formed that allows a spark to flash between the regions. This then is *lightning*, an electrical discharge between positive and negative regions in the environment. A lightning flash across the air heats the molecules around it. That heated volume expands and produces sound waves we hear as thunder.

The precursor to a lightning flash is a build-up of charge differences in the air. How does this build-up become established? The process is not totally locked down, but the build-up appears to derive from the processes that produce the thunderstorm. To form a thunderstorm, moist air and rapidly rising air currents are needed. When the moist air ascends to its condensation level, water droplets form and a cloud is born. When the rising currents are strong, the cloud droplets and larger raindrops that begin forming are jostled against one another. When the temperature of the rising air drops to below the freezing mark, much of the water forms ice crystals that rise ever higher in the cloud, and the rest forms ice pellets that collide and adhere.

During collisions among ice pellets

Mesotortuous segments form the path of a lightning bolt from cloud to ground.

Dramatic lightning storm over Boston, Massachusetts. (NATIONAL OCEANIC AND ATMOSPHERIC ADMINISTRATION/U.S. DEPARTMENT OF COMMERCE, NOAA HISTORIC NWS COLLECTION)

and ice crystals, some electrons are "captured" by one partner from the other, which separates the charges. This is similar to school science demonstrations that rub a glass or rubber rod through a piece of silk or fur to give it a charge. Over time, the charged particles can isolate themselves into concentrated pockets of like charges. Perhaps due to the size differential—and therefore the fall speed—between crystals and pellets, negative charges usually collect in the cloud's lower regions and positive charges gather in higher regions. The regions can be particularly well separated when an updraft in the cloud is near a downdraft, forming a *wind shear*. As a result of these interactions and transport, the cumulonimbus cloud forms a *dipole*, a body with concentrated charge of one sign in one region and the opposite in another. Since like charges repel, once the region in the cloud base gathers sufficient charge, it repels some of the charge in the electrically neutral ground beneath, producing a small region of opposite charge, usually positive, on the planetary surface beneath the cloud.

Forming charged pockets does not a lightning bolt make, however. In fact, the cumulonimbus must exceed a minimum size and particular makeup to become a thundercloud rather than just a rain cloud. Field studies suggest the cloud must contain more than 1 cubic kilometre (0.24 mi.³) or so of volume and be deep enough to cover the temperature range required for strong charge separation. Cumulonimbi extending 3 to 4 kilometres (1.8–2.5 mi.) horizontally and 5 to 8 kilometres (3–5 mi.) vertically are believed to define the minimum dimensions for a thundercloud.

With a typically charged thundercloud, the stage is set for developing lightning flashes. The lower portion of a thundercloud is usually negatively charged, while the higher regions are usually positively charged. Surrounding this cloud are other clouds that likely have charged regions and a formerly neutral Earth, where weakly charged regions have formed under the influence of the large charged clouds overhead. A lightning flash, however, will not occur until the resistance between charged regions is overcome.

In the typical situation of a cloud to ground strike, the potential between oppositely charged regions increases as the resistance of the intervening air begins to break down. The first signs of the breakdown are small streamers, channels of high-charged particle (*ion*) density, that propagate from the cloud to Earth. Streamers creep downward in small discrete steps (mesotortuous segments) about 50 metres (164 ft.) in length that may turn from a straight downward path to follow paths of least electrical resistance in the air. This entity, called a *stepped leader*, creates an ionized path that deposits electrical charge along the channel. Just before the stepped leader reaches the surface or an object extending from the surface, the potential difference between the end of the leader and the surface becomes extremely high.

At this time, a streamer emerges from the surface that intercepts the descending stepped leader just before it touches ground. When an uninterrupted path between cloud and surface is achieved, a return current of electricity surges up the ionized path at nearly the speed of light, initiating *a return stroke*. The return stroke releases a tremendous electrical energy surge and bright light, and a cloud-to-ground lightning flash has occurred. Once the pathway has been established, it may be used more than once to release the built-up potential. A typical lightning flash is composed of a series of strokes, averaging about four per flash. The duration of each stroke varies but generally is about 30 microseconds, thus the ensemble usually appears as one flash to our eyes. (Television screens flicker every 0.03 seconds.) In contrast, lightning strokes within a cloud are usually single strokes because inter-cloud and intra-cloud lightning do not produce a return stroke.

Different Strokes

Lightning from negatively charged areas of the cloud generally carries negative charge to Earth and is called a *negative flash*. These account for 95 percent of cloud-to-ground lightning strikes. A discharge from a positively charged cloud area produces a *positive flash*. Recent research from lightning detection networks suggests that positive flashes may be hotter and more likely to start forest or grass fires. They also are of longer duration and higher peak current.

Lightning strokes are quick but powerful. Lasting only a few milliseconds, they may carry thousands of watts of power and produce 30,000 amps. (A house uses only about 200 amps.) The average flash could light a 100-watt bulb for more than a month. Air within the lightning channel heats up to temperatures as high as 30,000°C (54,000°F) (about five times hotter than the solar surface). The typical lightning bolt is about 10 kilometres (0.6 mi.) long and 2 to 10 centimetres (1–4 in.) thick.

Besides the positive and negative flash differences, lightning has several common categories based upon where the strokes begin and end.

Cloud-to-ground lightning, the most dangerous and damaging form, jumps from a cloud to the ground or some object arising from the ground. Negative charge is brought to the earth by these bolts, which are characterized by downward pointing fork branches.

Ground-to-cloud lightning, the reverse pathway of the above, is not as common and can be distinguished from cloud-to-ground lightning because its "forks" point upward. These flashes generally transfer a net positive charge to the earth.

Intra-cloud lightning, the most common type of lightning discharge, is lightning arcing between regions in the same cloud. This form is often called *sheet lightning* because the stroke details lie hidden from view behind cloud elements and are seen only as a diffuse brightening, which may flicker within the cloud.

Inter-cloud lightning jumps from one cloud to another, the stroke bridging a gap of clear air between them.

Cloud-to-sky lightning is not commonly seen from the ground and was once considered rare; however, many recent observations from aircraft and space platforms have opened a whole new field of research. These flashes are often colored and have earned names such as red sprites, blue jets, and elves when they emanate from storm tops into the upper atmosphere.

An estimated eighteen hundred thunderstorms are raging around the globe as you read this. That amounts to 16 million storms per year. Globally, lightning flashes fifty to one hundred times every second—around 10 million hits a day. Recent observations suggest an

Lightning bolts dance across the city sky during a rare Pacific Coast thunderstorm on southern Vancouver Island.
(KEITH C. HEIDORN)

average of 25 million strokes of lightning leap from cloud to ground every year.

If you plan to observe lightning and thunder, please take proper safety precautions. More people are killed in North America each year by lightning than by tornadoes and hurricanes combined. Each year in Canada, lightning kills six to twelve people on average. It takes nearly one hundred lives and injures more than five hundred people per year in the United States.

When lightning strikes, an enclosed building is the safest place to be—open structures are not safe. If you can't be in a building, an enclosed vehicle with a hard top and the windows rolled up is an option—but remember, it's the metal casing around you that protects you, not the rubber tires. Don't touch the sides of the car.

Dispelling a Myth

An old folk belief states that lightning never strikes twice in the same place. Aside from the fact that a single flash may send two to twenty strokes along the same channel, lightning not only strikes twice, but many times. Toronto's CN Tower, at 555 metres (1,820 ft.) tall, is struck, on average, about seventy-five times per year, and the Empire State Building in New York City (444 metres [1,456 ft.]), twenty-one to twenty-five times annually. The reigning champion human lightning rod is U.S. Park Service Ranger Ray Sullivan, who was struck by lightning seven different times between 1942 and 1976, surviving, obviously, each strike.

Thunder: Lightning's Child

"First let me talk with this philosopher—
what is the cause of thunder?"
(William Shakespeare, *King Lear*, Act 3, Scene 4)

High on the list of most-recognized natural (non-human) sounds across the planet is weather's basso voice: *thunder*. Its name is incorporated into the name of one of the most common weather phenomena on Earth, the thunderstorm and its associated elements: the thunderhead, thundercloud, thunderbolt, thundershower, and thunderclap.

Early humans believed thunder was the voice of their gods. The roster of thunder gods includes Thor of the Scandinavians, Donar of the Germans, Zeus of the Greeks, Jupiter of the Romans, Taranis of the Celts, Perkunis of the Slavs, Indra of India, and Shango of the Nigerian Yoruba. Each was known to throw thunderbolts or bundles of lightning at Earth while their voices reverberated across the heavens.

A number of cultures recognize a thunderbird as being responsible for thunder and lightning. The power of the thunderbird is frequently found in the legends of Native American nations and African tribes. For example, the Bantu of Africa believed that thunder resulted from the beating of the wings of Umpundulo as the bird dove toward Earth.

Many early cultures believed thunder was an omen. For example, the Greeks thought thunder on the right was a good omen; however, the Romans regarded thunder on the left as favorable. Both agreed that thunder in the east was more favorable than thunder in the west—perhaps because, since weather generally moves from west to east, thunder to the east meant the tempest had passed.

Early Scientific Theory

Eventually, people began to realize that thunder (and lightning) had natural causes that could be explained through observation and logical deduction. The earliest known scientific theories of thunder were proposed by the Greeks Anaximander (circa 611–547 BC) and Anaximenes (circa 585–528 BC), both followers of the great Greek natural philosopher Thales (circa 600 BC). They both believed that air smashing against the clouds caused thunder and, as the air struggled its way through the clouds, it kindled a flame—lightning.

Anaxogoras (circa 499–427 BC) believed thunder resulted when fire flashing through clouds (lightning) was quenched by the water in the cloud. Democritus (circa 460–370 BC), on the other hand, suggested thunder and lightning were due to the unequal mixing of particles within clouds, which caused violent motions, the resulting sound of which was thunder.

Aristotle (384–322 BC), in his series of essays titled *Meteorologica* (written around 334 BC), wrote that thunder occurs when

> the dry exhalation that gets trapped when the air is in the process of colliding is forcibly ejected as the clouds condense and in its course strikes the surrounding clouds, and the noise caused by the impact is what we call thunder . . . similarly the windy exhalations in the clouds produces thunder when it strikes a dense cloud formation. Different kinds of sound are produced because of the lack of uniformity in the clouds and because hollows occur where their density is not continuous.

The Roman poet Lucretius, in his classic treatise on the universe, *Of the Nature of Things* (circa 60 BC), could not arrive at a single cause for thunder. As one cause, he supposed that:

> With thunder are shaken the blue deeps of heaven,
> Because the ethereal clouds, scudding aloft,
> Together clash, what time 'gainst one another
> The winds are battling. For never a sound there comes
> From out the serene regions of the sky;
> But wheresoever in a host more dense
> The clouds foregather, thence more often comes
> A crash with mighty rumbling.
> (from William Ellery Leonard's translation, Gutenberg
> e-text, Book IV)

Lucretius felt thunder could also be caused by the sound of hail and ice crashing together within the cloud or by the bursting of a cloud when the wind trapped inside escaped, like the popping of a balloon. He reasoned that lightning occurred simultaneously with thunder, the result of sparks thrown from the collision or bursting of clouds.

The Riddle Solved

All these early theories of thunder held sway for more than two thousand years. Even that astute observer of nature René Descartes suggested in 1637 that thunder arose when higher clouds descended onto lower ones, the sound of their collision reverberating in the air space between the cloud layers.

In the mid-seventeenth century, British physicist Robert Hooke deduced that the duration of thunder was dependent upon the distance between the lightning stroke and the observer. J. N. De L'Isle measured the time delay between observing the lightning flash and hearing the thunder. His 1738 paper noted that thunder was rarely heard from lightning discharges occurring more than 25 kilometres (15 mi.) away. Benjamin Franklin reasoned that if a spark produced in the laboratory produced a loud snap, lightning should also produce a sound. "How loud must be the crack of 10,000 acres of electrified cloud?" he wondered. L. C. Veenema observed nearly every thunderstorm in his area from 1895 to 1916 to determine how far away thunder could be heard. His studies confirmed that thunder generally could not be heard more than 25 kilometres (15 mi.) from the lightning flash, although in certain instances, it could be heard up to and beyond 100 kilometres (62 mi.).

By the end of the nineteenth century, the true relationship between lightning and thunder had finally been established, although the exact process of cause and effect was still being debated in scientific circles. These debates led to several common theories. The commonly held theory at the time proposed that thunder was produced when lightning, passing through the air, caused a vacuum to form. When this vacuum collapsed, the air rapidly rushed back in, producing a thunderous explosion. The *steam theory* of R. V. Reynolds (1903) proposed that the lightning bolt caused water in its path to be rapidly transformed to steam at enormous pressure, and it was the rapid expansion of that steam that caused the thundering sounds. R. S. Mershon had hypothesized, in 1870, that when lightning passed from one cloud to another, the electrical discharge decomposed water into its atomic components: hydrogen and oxygen. These elements then explosively recombined in the great heat of the lightning stroke. The explosion was the thunder.

J. A. Lyon in 1903 disagreed with the theories of both Mershon and Reynolds. He pointed out that electrical sparks produced in the laboratory through atmospheres devoid of water or explosive gases still produced a sound. Lyon's theory for thunder was similar to that proposed in 1888 by M. Hirn. In a *Scientific American* article that year, Hirn advanced the theory that thunder is due simply to air traversed by a flash of lightning being suddenly raised to a very high temperature. The gas thus heated and expanded produces a thunderclap.

Here at last we have a nearly complete answer to the thunder riddle. Atmospheric studies in the twentieth century broadened Hirn's work into the true nature of thunder.

The Modern Theory

Today, the origin of thunder is well known by atmospheric scientists. Since thunder is so intimately associated with lightning, much of what was learned about the location, shape, and orientation of lightning has been deduced using thunder's sound waves. We start the process with the two main ingredients: air and lightning. Having electrical resistance, air becomes heated when an electrical current, such as lightning, passes through it. Since each lightning flash has a temperature hotter than the surface of the sun, it superheats the air surrounding its path, resulting in a channel of gas at very high temperature and pressure. As the superheated gas rapidly expands into the surrounding air, first a shock wave and then a sound wave radiate from the lightning channel.

Lightning strokes are composed of a series of path segments. A lightning stroke surges between its beginning and end points in a series of steps, and individual lightning flashes may occur many times along that pathway in less than a second. Each surge in the lightning flash heats the air along the lightning channel, producing a series of acoustic waves. The loudness and duration of the resulting thunder is dependent on the strength of the lightning surge current. Research has found that those segments of the lightning channel between 5 and 100 metres (16 and 329 ft.) in length produce most of the audible pulses of thunder.

As mentioned earlier, the lightning bolt heats the surrounding air to as much as 30,000°C (54,000°F), causing a nearly instantaneous increase in pressure from ten to one hundred times normal atmospheric pressure. Thunder starts as a shock wave moving at speeds in excess of the speed of sound. This initial shock wave rapidly loses its energy to the surrounding air. When the energy that the shock wave received from the lightning stroke is expended, the wave "relaxes" and the pressure in the vicinity of the channel returns to normal levels. In relaxing, the shock wave produces an acoustic, or sound, wave, which radiates through the air perpendicular to the lightning segment that produced it.

Although less than 1 percent of the total energy in the initial shock wave is transformed into the acoustic wave (the remaining 99 percent is dissipated into heating the surrounding air), the total energy available for that sound wave is still extremely large. Therefore, thunder is one of nature's loudest sounds. A nearby thunderclap may reach a sound level of around 120 decibels, equivalent to being within 60 metres (197 ft.) of a jet aircraft during takeoff or about 1 metre (39 in.) of an automobile's horn. A chain saw is rated around 100 decibels. A sound of 140 decibels is painfully loud and can cause hearing damage.

The radius of the shock wave at the time of *relaxation* determines the characteristic frequency or *pitch* of the thunder from that stroke: the more powerful the lightning stroke, the wider the channel and the lower the pitch of the resulting thunder. Thunder generally has a pitch either between 15 and 40 hertz or 75 and 120 hertz. (The lowest note on a full keyboard piano has a tuned pitch of 66 hertz.)

Thunder is more than a simple, loud explosive sound following a lightning bolt. Thunder peals. It rolls and rumbles through the stormy sky. It cracks and claps. Thunder varies in duration and the distance over which it may be heard. What then accounts for this wide variety of sounds that we classify under the label "thunder"? Thunder travels through the lower atmosphere as an acoustic or sound wave moving at a speed of roughly 1,230 km/h (764 mph)—the speed of sound. The pitch, loudness, and form of thunder (crack, rumble, or boom) depend on the lightning flash that produced it, and the order in which the various sound waves from a lightning stroke reach the observer are all primarily determined by the lightning flash's shape and location. Sound waves are also modified by the atmosphere through which they travel.

Thunder sound waves originating from the lightning flash do not radiate with equal strength in all directions from the lightning chan-

nel. More than 80 percent of the acoustic energy is radiated into zones 30 degrees above and below the surface of the plane that perpendicularly bisects the spark. Since the average change in the direction or orientation between adjacent lightning segments of the lightning stroke is only about 16 degrees (smaller than the zone into which most of the acoustic energy is radiated), the thunderbolt's largest seg-

Sound waves radiate from lightning segment on a vertical bolt toward the observer.

Sound waves radiate from a lightning segment on a horizontal bolt toward the observer.

ments will emit their loudest sound in roughly the same direction. However, it is the degree of that orientation change between segments that determines whether thunder is heard as a sudden clap or a prolonged rumble.

Sound waves from all segments of the lightning stroke are produced almost simultaneously, over a time interval much less than a second in length. What variations we hear in a thunder peal result from the time required for the sound from different segments of the lightning bolt to reach our ears, the nearest segments being heard before the more distant. This time differential, coupled with the length and orientation of the lightning flash's larger segments, determines the unique character of each thunder peal we hear. For example, if the main channels of the lightning bolt are end-on to the listener, the thunder will be relatively quiet since most of the sound is being radiated perpendicular to the channel, away from the listener.

Since the sound is generated from portions of the bolt progressively farther from the listener, those sound waves that do reach the ear quickly combine to produce a prolonged soft roll or rumble. On the other hand, if the channel is broadside rather than end-on to the listener, most of the sound is directed toward the listener, and the sound waves from the various channel segments arrive almost simultaneously, resulting in a short, but loud, thunderclap. Since each lightning flash is composed of a number of large segments oriented in any number of ways relative to the listener, the thunder that one generally hears is a combination of claps and rumbles. Listeners separated by some distance will each perceive the thunder in a unique way.

Atmospheric Effects on Thunder

So far, we have assumed that the thunder sound wave has moved through an ideal atmosphere and flat terrain. Thunder does not, however, travel from the lightning channel to the receiver through a uniform atmosphere or always over an ideal, featureless terrain. Since the atmosphere varies in density and has winds blowing through it at various speeds and directions, thunder may be scattered, attenuated, refracted, or reflected, all of which alter the volume, pitch, and character of the thunder heard.

The scattering and attenuation processes in the atmosphere alter the total sound package reaching the listener, mostly by weakening the higher-pitched frequencies. Thus, by the time the thunder reaches a listener several kilometres from the lightning stroke, the predominant sound will be a low-pitched rumble. If the lightning flash happens to be of low energy—a situation that produces mostly higher-pitched sounds—no audible thunder will be heard except close to the lightning channel.

Air temperature and wind refract, or bend, the wave from a straight line path toward the listener. Since the temperature of the lower atmosphere usually decreases with height, and sound travels faster in warm than cold air, sound waves moving through the lower atmosphere curve upward. Thunder propagating from portions of the lightning bolt, therefore, may not be heard as refraction bends the sound wave up, causing it to pass over the head of the listener on the surface.

Wind can have two effects on thunder. First, it may increase or decrease the speed at which the sound wave moves through the air. Sound moves faster downwind than it does upwind. Second, the variation of wind with height may refract sound waves, similar to the effects of vertical temperature gradients. If wind increases with height, as is usually the case near Earth's surface, the sound wave refracts upward, bending away from the surface-based listener.

The combined effects of scattering, attenuation, and refraction limit the distances at which thunder may be heard by a ground-based observer. Although this distance varies with the temperature gradient, wind speed, and height of the lightning flash, thunder generally will not be heard farther than 10 to 25 kilometres (6–15 mi.) from the lightning bolt, in agreement with De L'Isle's and Veenema's much earlier findings. This effect is best observed in the phenomenon known as *heat lightning*, where lightning from distant thunderstorm cells is visible but from which no thunder is heard by the observer.

"Watching" Thunder

While the professional scientist may use sophisticated audio equipment to study lightning through the properties of its thunder signature, the amateur observer can make many interesting observations with no more equipment than a watch capable of measuring seconds and a critical ear.

The easiest observation is to estimate the distance the lightning stroke is from the observer. The distance between the lightning and the observer can be calculated by counting the number of seconds between the flash and the thunder and dividing by three for distances in kilometres or five for distances in miles.

With careful listening to thunder events, the shape, length, and path of a lightning bolt can often be estimated. By measuring the delay between the flash and 1) the first thunder heard, 2) the loudest clap, and 3) the final rumble, it is possible to estimate the distance to the nearest lightning branch, the main channel, and the farthest branch, respectively. Noting the total duration of the thunder allows calculation of the minimum length of the channel. Use the same method of calculation given above to convert the time interval in seconds to a distance or length in kilometres or miles.

Lightning striking the ground nearby gives a loud crack, at times preceded by a short rumble or a ripping noise that likely originates from a small branch of the main lightning channel. A flash made up of several different strokes may generate thunder such that each pulse is heard separately, sounding similar to a short burst of machine gun fire. Thunder heard as a loud boom rather than a clap is usually generated by a high-energy flash some distance from the observer. A crackling thunder racing from one side of the observer to the other results from an overhead, in-cloud lightning flash beginning on one side of the observer and ending on the other.

Thunder has the potential to arouse deep feelings. The sudden clap, particularly when unexpected, can scare even the calmest person or awaken one from deep sleep. A distant rumbling may bring sighs of relief from farmers needing rain for their crops or curses from the lips of golfers or picnickers, whose fun will soon be ended. But farmers may also curse the coming storm if it drops not nurturing rain on their crops, but hail.

Thirty-seven

Hail to Thee

Perhaps the weather gods had planned this in advance—or maybe they had just read my mind—but less than a week before I sat down to write this chapter, they presented me with an exhibition of hail. My present home turf is not blessed with many thundershowers or thunderstorms. Often in Victoria, thunderstorms are so short that by the time you can react, say "What's that?" and go to the window, the action is over. The storm that struck this day had a few rumbles of thunder—I saw no lightning myself. Summer hail is not as common to southeastern Vancouver Island (except in the mountains) as it is in other parts of the continent. It's more common here in the colder months, when moist air flows off the warmer Pacific and is thrust upward into colder air aloft by Vancouver Island's mountainous terrain. These hailstorms provide us with small, usually pea-sized, hail. But that June day, the weather gods treated me to three separate hail showers. The heaviest lasted about three minutes and produced dime-sized hail.

Hailstones

Hail is the prime warm-season species of frozen precipitation. When sliced through their center, hailstones reveal an onion-like layering of ice. These distinctive ice layers, alternating between opaque and clear, indicate the manner in which ice accumulated during different stages of the hailstone's growth. An opaque ice layer forms when the hailstone collects small, supercooled liquid water drops that freeze rapidly on impact, trapping air bubbles within the ice and giving it a milky texture. When larger water drops impact on a hailstone, the freezing is slower, allowing air bubbles to escape, forming clear ice.

The Canadian Meteorological Service and the U.S. National Weather Service have official terms for hailstone diameters, ranging from *pea-sized* to *softball-sized*. The British Meteorological Office uses a slightly different set of terms: *melon* rather than *grapefruit*, and *coconut* rather than *softball*, but *pea* is the same in both. Our *marble* size is their *mothball* (second-smallest category).

The largest hailstone officially documented for years fell in Coffeyville, Kansas, on 3 September 1970. It weighed 0.75 kilograms (1.63 lb.) and spanned 14.4 centimetres (5.7 in.). The previous record stone fell at Potter, Nebraska, on 6 July 1928, weighing about 0.68 kilograms (1.5 lb.) and measuring 13.7 centimetres (5.4 in.) in diameter. More recently, a new benchmark was set for the largest hailstone recovered in the United States. It hit Aurora, Nebraska, on 22 June 2003 with a record 17.8-centimetre (7.0-in.) diameter and a circumference of 37.6 centimetres (14.8 in.). The NOAA National Climate Extremes Committee, which is responsible for validating national records, has formally accepted the measurements. The stone failed to weigh in above the Coffeyville stone, though observers said it broke on impact and lost some of its mass.

Hailstone Formation

Hailstones generally begin forming on seeds of small frozen raindrops or soft ice particles known as *graupel*, which are hardened conglomerates of snowflakes. Hailstones also sometimes contain foreign matter, such as pebbles, leaves, twigs, and insects lofted into the storm cloud by strong updrafts.

Hail can be found in the middle and upper portions of almost all moderate to severe thunderstorms forming outside the tropics. A hailstone's size usually increases with the intensity of the storm cell from which it spawns. To form a hailstone the size of a golf ball, more than 10 billion supercooled droplets must be accumulated and must remain in the storm cloud for at least five to ten minutes. By comparison, 1 million or so droplets are needed to form a small raindrop. Large hail (more than 5 centimetres [2 in.] in diameter) forms mostly in supercell thunderstorms, which have strong updraft winds exceeding 80 km/h (50 mph).

In order for the frozen raindrops or graupel to grow into true hailstones, they must accumulate additional ice, a process called *accretion*. To do so, the hail embryo must spend time in cloud regions rich in supercooled water, a layer where temperatures are below 0°C (32°F). A layer of drier air in the mid-levels of the thunderstorm also assists in hail development. Because very moist air retains more heat, hailstones forming in an extremely moist atmosphere don't freeze properly and resemble soggy slush balls (one reason hail is rare in hurricanes). As these soft stones fall, they either disintegrate or melt well before reaching the ground.

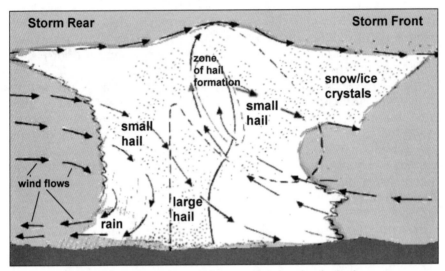

Hail formation regions within a cumulonimbus thundercloud.

Storm research has determined that hail is most likely to form when the height of the *wet-bulb zero level* falls between 2,200 and 2,800 metres (7,200 and 9,200 ft.). The wet-bulb zero level is an adjustment to the altitude at which the melting point of ice (0°C [32°F]) occurs and takes into account the added cooling effect generated by the evaporation of precipitation into the surrounding air. Precipitation falling through relatively dry air can cool the air further by evaporation of liquid water, lowering the height of the wet-bulb zero level below the altitude of the 0°C (32°F) temperature isotherm. It is important that the wet-bulb zero level falls within this narrow altitude range—the likelihood that a hailstone will form and then fall to the surface decreases when the level is outside this range. If the wet-bulb zero level is too high, most hailstones will melt before reaching the ground. A prime example of these conditions can be found in Florida. Although Florida is a hotbed for thunderstorm activity, the wet-bulb zero level is usually too high, thus hail fall is infrequent—most stones melt before reaching ground. If the wet-bulb zero level is too low—below 2,200 metres (7,200 ft.)—it usually indicates a relatively cold air layer at low altitudes that will inhibit the formation of strong updrafts within the thunderstorm needed to promote hailstone growth.

Hailstorms require strong updrafts, cold air layers, sufficient ice nuclei, and supercooled water to form. Thus, the squall line and supercell thunderstorms are the most frequent hail producers. Supercells have the greatest potential for damaging hail since they are physically taller clouds and have access to the atmosphere's colder levels. They are also longer lived, having a better chance of acquiring nuclei in sufficient quantities to form hail.

The Role of the Updraft

For a hail embryo to grow, it must remain in a layer of supercooled water for a length of time—the longer the residency, the larger the hailstone's potential size. Since Mother Earth has a strong attraction—gravity—for all things airborne, there must be a countering force to keep hailstones aloft. This is supplied by the "Storm King," in the form of strong updrafts that construct the great towers of cumulonimbus storm clouds. The updraft is not only the agent producing hail, it is also responsible for pushing cumulonimbus cloud tops high into the troposphere, creating the environment for lightning, thunder, and the various wind characteristics of the thunderstorm cell. For the smallest hail to form, an updraft of around 36 to 54 km/h (22–34 mph) is required. Larger stones, like those termed golf ball–sized, require updrafts of around 85 km/h (53 mph) to form. Softball-sized hail involves updrafts exceeding 160 km/h (99 mph).

Fifty years ago, the widely held theory of hailstorm formation envisioned the growing hailstones riding a roller coaster of thunderstorm updrafts and downdrafts before falling earthward, when their weight finally exceeded the updrafts' lifting capabilities. We now believe that this is but one possible mechanism for hail formation. Hail need not ride a cumulonimbus elevator, but may increase in size by falling slowly through a layer rich in supercooled water. Such hailstones are often characterized by having little layering of the ice within. In the thunderstorm interior's rough and tumble environment, hailstones of various sizes may collide. The force of collision may break the stones into smaller ones, or it may weld stones together.

Eventually, each hailstone breaks free from its aerie and plunges to Earth. Its fall may be the result of an increase to a weight that the storm cell's updrafts can't support, or it may be caught in a downdraft and hurled earthward. Large hailstones fall at speeds faster than 160 km/h (99 mph). Not all hailstones survive the downward trip, however. Some are broken through collisions into smaller chunks that then melt before reaching the surface, while others just melt as they traverse the warmer (above freezing) air below. An estimated 40 to 70 percent of hailstones formed within a cumulonimbus nursery melt before touching ground.

When hail falls, it's along paths known as *hail swaths*. These can be rather small, only a hectare or so in area, or as large as 16 kilometres (10 mi.) by 160 kilometres (99 mi.). Hail swaths can pile hail

Large hailstone. Ruler shows radius of approximately 15 cm (6 in.), the size of a grapefruit.
(NOAA PHOTO LIBRARY, NOAA CENTRAL LIBRARY; OAR/ERL/NATIONAL SEVERE STORMS LABORATORY)

so deep that it must be removed with a snowplow. In August 1980, "hail drifts" were reported 2 metres (6.6 ft.) deep in Orient, Iowa. Severe hail swaths can devastate one field of crops while leaving a neighbor's untouched.

Hail Climatology

The regions in Canada with the most hail days are the central and eastern prairie provinces (parts of Alberta, Saskatchewan, and Manitoba), south-central British Columbia, and southwestern Ontario. The highest annual hail frequencies in Canada (for the 1977–93 data period) were between three and nine hail days during the warm months of May to September. Canada's "hailstorm alleys" are in British Columbia's interior and Alberta, with five to nine hail days per season. In these regions, hail most frequently forms between noon and evening, usually from May to July. In the United States, hailstorms are most common on the plains, especially east of the Rockies. Hail Alley—the most hail-struck region in the U.S.—is found where Colorado, Nebraska, and Wyoming meet, averaging between seven and nine hail days per year. Hail in this part of the country is most likely to fall late in the afternoon during May and June. Other parts of the world where damaging hailstorms are frequent include China, Russia, India, and northern Italy.

Hail Damage

Hailstorms are among the most damaging weather events each year in Canada and the United States. Costs are conservatively estimated at $100 million per year in Canada. The most costly Canadian hailstorm occurred on 7 September 1991 in Calgary, causing an estimated $360 million in insured damage and $450 million in estimated total damage. Damage in the United States approaches US$1 billion annually.

The most destructive hailstorm recorded in North America occurred around Dallas on 5 May 1995, causing an estimated damage of US$2 billion and injuring 510 people. The previous record holder was a Denver hailstorm whose softball-sized hail caused US$625 million in property damage on 11 July 1990. Much of the damage from hail is inflicted on crops, earning hail the title of the *White Plague* within the agriculture industry. Damage to vehicles, buildings—particularly roofs and glass—and landscaping leads the list of damage outside the agricultural sector.

Hail of any size can cause injury to humans and animals, and can even be fatal. The last reported hail death in North America, an infant, occurred in August 1979 at Fort Collins, Colorado. The most deadly hailstorm on record occurred in India on 30 April 1888, killing 246 people and 1,600 domesticated animals.

A severe thunderstorm deposits large hail on a town in the American Midwest. (NOAA PHOTO LIBRARY, NOAA CENTRAL LIBRARY; OAR/ERL/NATIONAL SEVERE STORMS LABORATORY)

Thirty-eight

Dust Devils

Peering across the sun-drenched expanse of the wide flat land of the Canadian prairies, I saw a rising dusty column dancing across the barren soil. With no clouds of any note in the surrounding sky, I knew immediately what I was seeing: a *dust devil*. In their most impressive state, dust devils have often been confused with small tornadoes, their rising shafts of air tossing dust and small objects around the rotating columns. While dust devils have a similar basic structure to tornadoes, they have one very major difference. Dust devils form from the ground up and are rarely connected to an overhead cloud of any size. True tornadoes always descend from a large, energetic cumulonimbus cloud.

As a natural part of everyday weather, dust devils rank second to turbulent eddies as the most common form of natural atmospheric vortices on Earth. Likely more form than are seen, particularly over surfaces that do not have much dust or other light materials to gather into the vortex. Also called the whirlwind, desert whirlwind, dancing dervish, desert devil, sand devil, Willy-Willy, and the Cockeyed Bob, an estimated ten dust devils roam Earth at any given moment.

In most cases, dust devils begin on hot, sunny days over dry terrain. These conditions amplify the factors that cause rapidly rising air columns to form. Under intense solar heating, the surface temperature of dry patches of land soars, often as high as 55°C (131°F). The surface heat is then quickly transferred to the overlying layer of air. The resulting hot parcel becomes much less dense than the surrounding air, and its buoyancy generates a hot air column, rising as invisible chimneys.

As this air rises, something gives it a little spin—perhaps the breeze aloft, perhaps a passing car at ground level, or perhaps the influx of cooler air descending to replace the rising parcel. If the whirling updraft column continues to intensify, the vortex tightens and rotates faster, like a spinning figure skater bringing in her arms. When the invisible winds pick up loose surface dust and soil, a dust devil is born. Since they form from individual rising columns, multiple vortices may skip across the landscape in pairs or in swarms. Due to their relatively small size and rather brief existence, dust devils are not influenced by the Coriolis parameter that sends large-scale weather patterns twisting, and thus may rotate either clockwise or counterclockwise, depending only on the direction arising from their initial twist. It is not particularly unusual to see a pair of dust devils, each spinning in a different direction as they march side by side across the landscape.

The majority of dust devils are small, usually less than 1 metre (3.3 ft.) in diameter, and rather short-lived, lasting a couple of minutes or less. The larger and longer-lived devils, however, garner the most attention. The largest can expand to 300 metres (1,000 ft.) wide and hundreds high, but most often they are only a few metres in diameter and less than thirty metres (100 ft.) tall. Their life spans are typically fifteen minutes, though a well-established devil can persist for an hour or more. Average dust devil winds swirl with wind speeds of 25 to 40 km/h (15–25 mph), but strong devils have been clocked at 152 km/h (94 mph). They can amble along the ground at 15 km/h (9 mph) or race across the terrain as fast as 100 km/h (62 mph). Larger ones have the potential to cause damage with their gusty winds, mostly through the impacts of their swirling material. In a typical desert dust devil, two dump truck loads (around 100 kilograms [220 lb.]) of dust, sand, or soil may be carried aloft. Large ones can contain enough suspended material to fill six dump trucks. Though dust, soil, and sand are the most common cargos, any light material, such as straw or leaves, can be swirled into circulation.

We generally think of dust devils arising over large arid expanses, such as deserts or drought-stricken farmland, a common device used in motion pictures and documentaries to signify heat and dry landscapes. However, even a large parking lot can generate enough heat to start a sizable devil spinning upward, and at times, these go unnoticed due to the sparsity of dust to make the rising column visible.

Dust devils aren't confined to our planet. Images from the *Mars Global Surveyor* in May 1999 caught dust devils 8 kilometres (5 mi.) tall cavorting over the dusty Red Planet. Dust devils have been observed leaving 15-metre (49-ft.)-wide tracks extending for several kilometres on the red surface.

Thirty-nine

Lammas: Summer's Turning Point

With the start of August, we reach the third cross-quarter day of the year, and the one not celebrated as much today as it was in Old Europe. Most Canadian provinces have established an early August holiday, but it has no real identity. Americans just bemoan the fact that no holidays fall between the Fourth of July and Labor Day. For the old agrarian cultures, however, this was a time of high celebrations that go by many names, including *Lughnasad* (Celtic) and *Lughnassadh* (Irish). I prefer the more easily pronounced Anglo-Saxon name: Lammas.

From a technical viewpoint, the third cross-quarter day marks the crossing from solar summer to solar autumn. The sunniest stretch of the year now falls behind, and day length starts to decline more rapidly as we head toward the equinox. While the sun may be waning in strength, in many areas, the warmest days of the year are just past or yet to come. Here on the Pacific coast, August is often the warmest month as the waters of the northern Pacific have just about reached their peak annual warmth.

Summer's turning point comes at a time when nature begins to show the early fruits of solar summer's labors, and this is manifested in agricultural production. To the agrarian societies of medieval Europe, early August signaled the beginning of the harvest season, the time when the first grains were harvested and many fruits and vegetables ripened, ready for picking. A quarter of the annual solar wheel has now turned since the celebration of Beltane, the time of planting crops and vegetable gardens.

In pagan cultures, Lammastide was the time to honor the mighty Sun God and the gods of the grain by ritualistically sacrificing the first grains to ensure the continuity of life. Lammas was, to the Celts, one of four Great Fire festivals held on the cross-quarter days. During Lammas, the custom of lighting bonfires was intended to add strength to the powers of the waning sun. Afterward, firebrands were kept in the home through the winter as protection against storms, lightning, and fires caused by lightning. Across North America, the many small town or county fairs held each year echo the Lammas tradition. Their agricultural competitions and midway games resemble the ancient European festivals at which people gathered to pay homage to the land and the fruits of their labor and to take part in community reverie.

With the beginning of solar autumn at Lammastide, the sun enters its "old age," its golden months. The heat of summer lingers a little longer, perhaps even bringing in the hot "dog days" of August. The ripening grains are followed by the ripening of the fruits of tree and vine. It is a perfect time to give thanks to Mother Earth for her bounty and beauty. Early August is a time to rejoice and be festive, a time to honor those among us who still know how to reap the harvest and connect us with our ancestors.

Dust in the Wind

As any housekeeper can tell you, dust in the air presents a never-ending scourge. Dust from industrial and other human activities are major air-quality concerns across North America. For example, road traffic, including particles generated from tire and brake wear, deposits millions of tonnes of dust per year over our larger cities. Dust can also be an important component of precipitation formation, the particles serving as condensation nuclei for water vapor, but in too high concentrations, dust particles may also inhibit precipitation by partitioning the available water vapor too thinly among the droplets.[5]

Recent research by Daniel Rosenfeld and other scientists from Israel's Hebrew University and the Weizmann Institute has revealed that excessive dust may actually amplify the process of desert formation. Human activities such as grazing and agricultural cultivation, which expose and disrupt the topsoil, increase the dust burden in the air. As more dust reaches the clouds, they in turn produce smaller raindrops that grow more slowly, are more likely to evaporate while falling, and thus yield less rainfall over their lifetime. This process can exacerbate drought conditions and contribute to the further desertification of the region. Dust plays an important role in the planetary heat balance by intercepting sunlight before it reaches the ground. By shading Earth's surface from the sun's radiation, dust aerosols have the same cooling effect as rain clouds.

Blowing Dust and Dust Storms
Many small-scale dust events are caused by localized high winds associated with the passage of weather fronts and systems, or by

5 For purposes of this discussion, *dust* refers to fine, solid earthen particles—soil, silt, or sand—not to industrial particle releases.

strong convective storm events such as a thunderstorm outflow winds or downbursts. In many areas of the American western and plains states, blowing dust (and sand) events are becoming common hazards due to severe reductions in visibility, particularly to highway travel. The U.S. National Weather Service will issue a *blowing dust (blowing sand) advisory* whenever the presence of local windblown dust or sand reduces visibility to 400 metres (0.25 mi.) or less. A *blowing dust (blowing sand) warning* is issued when blowing dust or sand reduces visibility to near zero.

On those occasions when blowing dust covers a large area, the event is termed a *dust storm*. Large-scale dust storms are often characterized by sustained high winds at the surface that are associated with synoptic-scale windstorms. Dust storms can last from three to four hours to several days and often occur during the late winter and early spring, when extreme pressure gradients produce strong winds over bare soil. American observational practices define a dust storm as occurring when horizontal visibility is less than 0.63 miles (1 km) but not less than 0.31 miles (500 m). When visibility falls below the 0.31-mile limit, the event is termed a severe dust storm.

How Dust Moves

Dust storms and "black blizzards" are the ultimate expression of blowing dust. Essentially, a dust storm is born when strong winds traverse dry, arid land with little vegetation and elevate tiny particles of sand, dust, and other debris skyward. To see the origin of blowing dust, we must get down on our knees and take a closer look at the surface of Earth. It all starts with a process known as *aeolian* (or *eolian*) transportation, the process whereby the wind picks up and transports surface dust particles. That transport may occur through *suspension, saltation,* or *creep* of the dust particles.

The smallest dust particles are held in the atmosphere through suspension. Suspension occurs when surface materials are lifted into the air, and upward air currents are strong enough to support the weight of the particles and hold them aloft indefinitely. Typical wind speeds near Earth's surface will suspend particles with diameters less than 0.2 millimetres (0.008 in.). Severe windstorms can hold large particles, caught within turbulent eddies, aloft for some time and push them to high altitudes, which enhances their travel distance. Under strong wind conditions, suspended dust particles may be lifted thousands of metres upward and carried thousands of kilometres downwind, held in suspension by turbulent eddies and updrafts.

Saltation (Latin for "leaping") moves small particles forward through a series of jumps or skips, like a game of "leap-tag." Saltation,

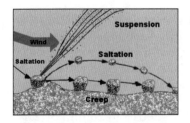

Surface dust is moved in the wind by suspension, saltation, and creep.

under light to moderate wind speeds, normally lifts particles no larger than 1 centimetre (0.4 in.) in diameter above the ground about 1 to 2 metres (3.3–6.6 ft.). These fine soil particles are lifted into the air, drifting across the surface approximately four times farther downwind than the height they attain. These dust particles are suspended only briefly as they are too heavy to remain airborne long. A saltating particle may hit another particle as it returns to Earth, and that particle will then jump up and forward to continue the saltation process.

Saltation is more or less a continuous process at high wind speeds. From a distance, a field of saltating particles appear as if they are constantly suspended, creating a fuzzy layer next to the ground. When saltating particles hit larger particles too heavy to hop, they nudge the larger grains, which may be up to six times larger than saltating particles, slightly and slowly forward, a sliding and rolling movement known as creep. Creep usually requires a wind speed exceeding 16 km/h (10 mph).

Aeolian transport is an important process for soil erosion, dune formation and alteration, and redeposition of soil particles. The great deposits of windblown soil around the world are known as *loess*, much of which originated from the debris left by glacial action during the last ice age. The thickest known loess deposit—335 metres (1,098 ft.)—is on the Loess Plateau in China. Deep accumulations of loess range generally from 20 to 30 metres (65–98 ft.) thick in Europe and the Americas.

If the lofted dust and debris form a large cloud or wall of dust that sweeps forward with the wind across large areas, we have a dust storm. Dust storms can be caused by various meteorological events, generally falling into one of two main categories: *convective events* and *large-scale non-convective events*.

Convective-event dust storms are usually associated with thunderstorm outflows or downbursts and form during the late afternoon in the spring and summer, when daytime heating is at its peak. They are far less predictable than the large-scale non-convective events. The duration of convective-event dust storms depends on the processes that form them. One caused by a microburst may last only a few intense seconds, while one caused by a downburst may last several minutes. Dust storms caused by winds associated with a dry squall line may last for hours.

Large-scale non-convective events arise from sustained high winds at the surface that are often associated with cyclonic (low-pressure cell) windstorms. They can last from three to four hours, even several days, depending on the overriding weather situation. Large-scale non-convective dust storms generally occur in late winter or early spring, when extreme pressure gradients produce high wind speeds and surfaces are often dry and bare. About two-thirds of dust storms are caused by strong winds associated with the passage of fronts and troughs or the down-mixing of upper level winds.

Global Dust Storms

When blowing winds are strong and persistent, dust from large deserts or bare agricultural land can ascend into the upper atmosphere and hitch a ride on the global circulation to distant lands. Over the past decade, China has been the source of major dust storms whose particles have crossed vast distances. In April 2001, an Asian dust storm kicked up a million-tonne dust cloud from the Gobi and Takla Makan deserts in Mongolia and China. Blown by global air currents, the leading edge of this cloud reached the North American Pacific coast on 12 April. Two days later, it had crossed the continent and moved off the eastern shoreline into the Atlantic.

A dust storm approaches Spearman, Texas, 14 April 1935, during the great "Dust Bowl" years.
(NATIONAL OCEANIC AND ATMOSPHERIC ADMINISTRATION/U.S. DEPARTMENT OF COMMERCE,
NOAA HISTORIC NWS COLLECTION)

Exactly a year later, in April 2002, another great dust storm sent a cloud out of China and over the Pacific toward the North American west coast. The telltale yellow stain from this soil tinted the air enough to make it visible to satellites overhead. From the ground below, the overhead sky appeared hazy white due to the scattering of sunlight by the dust particles.

The Sahara Desert has long been known for its dust and sand storms, and many special names have been given to these unwelcome episodes, including the *haboob*, the *harmattan*, and the *simoom*. Sahara dust is not only transported by the winds across northern Africa and into Europe, it can reach westward to Caribbean and West Indies waters, too. During many past summers, the U.S. National Weather Service Office in San Juan, Puerto Rico, has issued air pollution alerts due to the dust originating from the Sahara.

Local dust storms and rising warm air initially lift the Saharan sand to altitudes of around 4,500 metres (14,750 ft.), where upper-level subtropical winds carry the dust westward. While the larger particles fall out as the dust crosses the Atlantic, smaller particles continue into the northeastern Caribbean Sea, at times even reaching southern Florida. These dust clouds cause skies to turn hazy and reduce visibility, but the dust likely has more impacts than altering sky color and visibility. Researchers have linked Sahara dust to red tides, fish kills, and coral destruction, and now believe that such dust plays an important role in fertilizing the great Amazon rain forest.

Forty-one

Distant Thunderstorms at Night

Although the sun had set on a hot Illinois summer's day, to everyone's chagrin it ushered in a hot summer's evening. I noticed the northwestern sky glow with random flashes of light: the usual sign of an approaching thunderstorm—but no thunder followed, although the flashing sky continued for several hours. We had only a teasing episode of heat lightning, visible flashes of distant bolts lacking the slightest hint of thunder. This scene played out many times in my youth, growing up in northeastern Illinois. Today, living on Vancouver Island, thundershowers are few and far between and are not even seen in distant skies. I had not thought about such incidents for some time. Recently, I received several questions on this weather phenomenon and decided it was time to revisit *heat lightning.*

Folk weather mythology suggests heat lightning is caused by hot air expanding until it sparks on sultry summer nights—an incorrect hypothesis. On the other hand, there are many credible sources stating that heat lightning does not exist. This is mostly a matter of semantics. Heat lightning, you see, is *not* a unique form of lightning, but normal thunderstorm lightning that flashes too far away from observers for its thunder to be heard. It is most commonly called sheet lightning, a standard lightning bolt whose light is diffusely reflected off cumulonimbus cloud towers and thus has lost its distinctive bolt pattern.

During sultry weather, scattered, rather short-lived thunderstorms may pop up across a region, driven by excessive heat and humidity. Some travel overhead, bringing heat-relieving rain and cooling winds. Others pass a moderate distance away but not too far. Their lightning takes on many visual forms—forked, sheet, ribbon—and rumbles of thunder are heard, but no rain falls within view. Finally, some pass a long distance from us, their presence noted by only their towering cumulonimbus and flashes of yellow-tinted lightning (the blue wavelengths are scattered out of the bolt's color). Since scattered thunderstorms do not produce a dense, sky-filling cloud deck, we are presented with a long line of sight through the mostly clear air, often extending to the horizon. Therefore, we can sometimes see the upper levels of thunderstorms even at distances located below the horizon.

Within those distant thunderstorms, lightning bolts flash, and their light can be seen from as far away as 160 kilometres (100 mi.), depending on the bolt's height within the cloud, the air's clarity between us and the bolt, and our elevation above the ground. Thunder, in comparison, has a much shorter detection range—usually less than 25 kilometres (15 mi.) in a quiet rural setting and less than 8 kilometres (5 mi.) in a noisy city. Several factors limit sound's range of detection. Most important is the scattering and attenuation of sound by air molecules, particularly diminishing the higher-pitched sound frequencies. Thus, by the time thunder reaches a listener several kilometres from the lightning stroke, the predominant sound is a low-pitched rumble. Farther away, even the rumble ceases to be heard.

Vertical changes in temperature and wind speed through the lower atmosphere also affect the propagation of thunder by refracting sound waves from a straight line path toward the listener. Thus, thunder from high altitudes may bend away from the ground and never be heard. Even nearby thunder can have its sound waves bent away from an observer, and in certain situations, bend back earthward farther away, creating a dead zone for the storm's sounds. The combined effects of scattering, attenuation, refraction, and, in some cases, reflection, generally limit the distances at which thunder is heard to 10 to 25 kilometres (6–15 mi.) from the lightning bolt.

I'll conclude with a little folk weather wisdom. An old folk saying goes "Yeller gal, Yeller gal, flashing through the night, / Summer storm'll pass you, unless the lightnin's white."

And so it is with heat lightning, that "yeller" (yellow) gal of hot sultry nights.

It's Not Just the Heat

The air hangs heavy today. It has been one of those incredibly hot, humid days when you would love to remove your skin, but instead you wander tortoise-slow around the house, wearing the least clothing the law—or your family—will allow.

Throughout most of eastern North America, the Bermuda High—at its peak summer strength—continues to pump tropical air from the Gulf of Mexico north into Canada. No breeze stirs, nor rain falls, although the sky is so laden with moisture that perspiration clings to the walker as if it had found a long-lost relative. The temperature is 30°C (86°F) and rising, the relative humidity reads 75 percent, and it isn't even noon.

Once established, the heat wave intensifies. The stifling northward flow of air from the Gulf of Mexico gains heat as it passes over the hot land. Residents of the eastern half of North America languidly await the cool, dry air from the Canadian north. Later in the afternoon, thunderstorms spawned by the heat and humidity give temporary relief, but soon after, the added moisture drives up the humidity, making it even stickier. The high humidity also helps the air retain heat after the sun goes down, making nights unfit for sleep.

In years past, farm residents and workers, young and old, would gather around the closest rain barrel to sit and soak their feet in the cooling waters or jump in a nearby pond for relief. Today, we huddle close to the air conditioner and find excuses for not venturing outdoors, or we rely on fans and swimming pools to help alleviate the discomfort of sweltering summer heat.

In addition to the effects of heat, personal comfort and health are strongly affected by humidity—the moisture content of the air. The combination of heat and high humidity may cause discomfort, heatstroke, or even death to humans and animals. In the United States,

179

no weather condition killed more people annually than high heat and humidity over the decade 1994–2003: an average of 237 per year.

How the Body Reacts to Excess Heat

When blood is heated above 37°C (98.6°F), the human body attempts to lose the extra heat by expelling water through the skin and sweat glands, altering the rate and depth of blood circulation, and, as a last resort, by panting. As body temperature rises, the heart rate increases and blood vessels dilate to increase blood flow from the body's core to the skin's surface. There, blood is circulated through tiny capillaries threading around the upper layers of skin, and some excess heat drains off into the hopefully cooler atmosphere. At the same time, water moves from the blood through the skin, a process we refer to as *perspiration* or sweating. About 90 percent of the body's heat is lost through the skin, and most of that is lost through perspiration. When the air temperature is greater than 37°C (98.6°F), heat can be lost only through sweating.

Sweating by itself, however, does nothing to cool the body unless the water is removed by evaporation. In order to evaporate sweat from the body, heat energy is required to change liquid water into vapor: 540 calories of heat energy per gram of water to be exact. That heat of vaporization is contributed by the body to the water—that is where the cooling comes from. However, some water vapor also condenses back onto the body, returning that heat in the process. If the rate of evaporation exceeds the rate of condensation, the body will cool. The concentration of water vapor in the air, the humidity, is a prime factor in determining the degree of evaporative cooling.

Those who inhabit areas affected by both heat and humidity often wonder about the relative comfort of desert heat, which often exceeds 40°C (104°F). The answer lies in the humidity. In a hot, dry environment, more heat can be lost through evaporation than regained through condensation, and the body cools. When humidity is high, however, much of the heat lost is countered by an almost equal heat gain; thus the cooling of the body is minimal, leading to overheating. Overheating can cause discomfort at the very least and death at the very worst. Continued loss of water and a variety of dissolved chemicals such as sodium chloride (salt) from the body, if not replenished, can cause dehydration and internal chemical imbalances. Dehydration depletes the body of water needed for sweating and thickens the blood, requiring more pressure to pump it through the body, straining the heart and blood vessels. Increased heart rate and blood flow may quickly harm or kill those with heart or circulatory diseases. Research on the effects of heat and humidity on humans has shown that the severity of heat disorders increases with age. Conditions that cause heat cramps in a sixteen-year-old may cause heat exhaustion in a forty-year-old and heatstroke in someone over sixty.

Giving Discomfort a Number

Since humidity affects the body's ability to cool itself, *biometeorologists*—scientists who study the relationship between weather and life—have looked for ways in which to combine air temperature and humidity. The dangers posed by heat and humidity have led to the development of various heat-humidity discomfort indices. In order to alert residents to the combined dangers of heat and humidity, Canadian and American weather services issue heat warnings similar to winter's windchill warnings. The Canadian index is called the *Humidex*; in the United States, it is the *Heat Index Apparent Temperature*, more commonly known as the Heat Index. They are based on slightly different mathematical combinations of temperature and relative humidity.

Since these indices are, to some degree, subjective, the level of discomfort or danger depends on a person's age, health and physical condition, the type and amount of clothing worn, and activity level. Besides the temperature and humidity, weather conditions such as amount of sunshine and wind speed also affect the "feel" of temperature and humidity.

Humidex

Humidex combines the temperature and humidity into one number intended to reflect perceived temperature. The Humidex value reported must be equal to or greater than the air temperature. If the calculated value is less, the Humidex value and air temperature are the same.

Perceived Conditions for Humidex Values

Humidex	Comfort level
Less than 29°C	No discomfort
30 to 39°C	Some discomfort
40 to 45°C	Great discomfort; exertion should be minimal
45 to 53°C	Dangerous conditions; exertion should be avoided
Above 54°C	Heatstroke imminent

The Heat Index

In the United States, forecasters use the Heat Index Apparent Temperature, more commonly known as the Heat Index, as their accepted measure of thermal discomfort. It is a simplification of an index developed in 1979 by R. I. Steadman. Originally, Steadman included twenty factors in his index, but the Heat Index is calculated only from temperature (in degrees Fahrenheit) and relative humidity (in percent) using a complex formula.

Relationship between the Heat Index and Heat Disorders

Heat index	Health effects
80 to 90°F	Fatigue possible with prolonged exposure and/or physical activity
90 to 105°F	Heat cramps and heat exhaustion possible with prolonged exposure and/or physical activity
105 to 130°F	Heat cramps or heat exhaustion likely and heatstroke possible with prolonged exposure and/or physical activity
130°F or higher	Heatstroke highly likely with continued exposure

Forty-three

Hot Town, Summer in the City

"Hot town, summer in the city," the opening line of a 1966 hit by the Lovin' Spoonful, describes a familiar condition to urban residents. To meteorologists, this condition is known as the *urban heat island* and distinguishes a city's climate from the surrounding rural areas through all seasons. Most of us are aware of its presence in the colder months, when local weather forecasters announce, "Low tonight, four degrees, cooler in the suburbs" or "Risk of frost in the outlying areas."

The urban heat island is an area, usually centered on the city core, of air temperatures warmer than those in the surrounding countryside. On an infrared satellite picture, cities show up as islands of heat amid a sea of cooler temperatures. The warmest temperatures are usually located at the city center, with temperatures dropping off toward the city edge. A further drop is usually apparent in the surrounding suburbs, with minimum temperatures found in rural surroundings. The diagram on page 184 shows an idealized schematic of the urban heat island structure, but in reality, the picture is not always as smooth. As infrared satellite images show, pockets of warm air and cool air (usually parks) make up the thermal topography of the heat island.

When compared to the encircling countryside, the urban heat island has significant impacts on a city's climate. It reduces heating needs, snowfall amounts, and snow cover duration during the winter in many cities. It alters the wind speed and direction (in conjunction with other urban elements), cloud cover, precipitation, and air quality, year-round. But it is during the summer that we most feel the impact of the urban heat island, when it makes walking on the sidewalks feel like walking on hot coals.

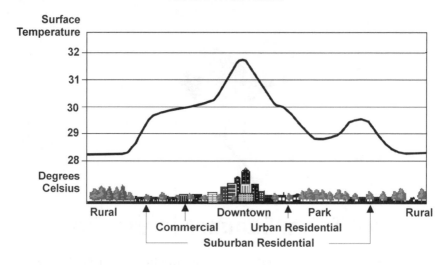

Midday surface temperature variation across a city showing urban heat island effect.

First Recognition

That the environment of cities differs from the surrounding country-side has been recognized for millennia. At first, it was poor air quality—mostly smoke from heating fires or industrial processes—that characterized the city environment. In 1818, when Luke Howard published the first scientific study of urban climates, *The Climate of London*, more details of the character of urban climate emerged.

Howard, who we met earlier as the man who named the clouds, was the first to propose that the urban center was warmer than the surrounding countryside at night and have the data to back up the finding. In his book, beneath a table of a nine-year comparison between temperature readings in London and the surrounding country, he commented, "Night is 3.70 [F]° warmer and day 0.34 [F]° cooler in the city than in the country." He attributed this difference to the extensive use of fuel in the city, with the daytime cooling a result of the urban smoke haze reducing the penetration of sunlight.

Since Howard's groundbreaking report on London, that city and many others around the world have been studied extensively to determine the full urban impacts on local and regional climate. While individual situations vary in detail, the general picture indicates that cities are noticeably and significantly warmer than the surrounding countryside, particularly at night. Furthermore, the urban heat island effect increases as cities grow in size and population. But temperatures are not all that are affected. Average precipitation, humidity, and wind speeds are also altered by an urban area. So, too, are specific weather events such as thunderstorms.

The Storms of "Hot-Lanta"

Besides altering climate normals such as mean temperature and precipitation, many cities are capable of altering their weather as well. Once a subject of hot debate, the evidence for this today is more convincing. The 1996 Summer Olympics in Atlanta, Georgia, provided more than just fun and athletic competition; they also provided an interesting confirmation of what some had believed prior to the Games, that Atlanta was experiencing more thunderstorms than it had a few decades earlier.

A dense weather observation network across Atlanta was established with a team of meteorologists to provide Olympics officials and athletes with up-to-date weather conditions at the many venues and quick advisories of any weather hazards that could arise. After the Games, meteorologists from California's San Jose State University studied these weather records and arrived at a startling conclusion: Atlanta generated its own thunderstorms!

To see the full picture, we must go back to the early 1970s, when Metropolitan Atlanta began a major growth spurt. From 1970 to 1980, the population grew by 27 percent and then exceeded that benchmark the following decade with a 33 percent increase. Suburbs doubled in size, and nearly 142,000 hectares (350,000 acres) of forest were cleared over the thirteen metropolitan counties. As dark roofs and pavement replaced green vegetation, the Atlanta heat island grew rapidly, at times recording an urban-rural temperature contrast as high as 5.6 C° (10 F°).

As Atlanta expanded, local weather observers began to notice an increase in the frequency of non-frontal, or air mass, thunderstorms and their attendant precipitation. Most striking, however, was a shift in the time of occurrence of these thunderstorm events. Non-frontal thunderstorms usually occur from mid- to late afternoon, following the daily peak in solar heating and maximum air temperature needed for their formation. But Atlanta weather observers were frequently recording early-morning thunder, a time when such storms should have been infrequent. Using data from the Olympic network, the San Jose State research team discovered convincing evidence that Greater Atlanta urban heat generation was indeed triggering thunderstorms south of the city. City heat was replacing solar heat as the initiation mechanism for cumulus growth and storm development, and its greatest contrast with surrounding regional conditions occurred in the early-morning hours.

It is most unlikely that Atlanta is a special case, and other urban areas in the United States and around the globe should also trigger *urban thunderstorms*. Researchers and public safety specialists have thus become concerned that urban-generated storms could trigger flash flooding and increased lightning and wind damage.

Urban Heat Hazards

Summer in the city can be especially discomforting; in fact, it can become extremely deadly. According to U.S. National Weather Service statistics, American heat-related deaths from 1994 to 2003 exceeded fatalities from any severe weather (i.e., storm) category. Averaged annually over that ten-year period, hurricanes killed 18 people, tornadoes claimed 58 lives, and lightning strikes and floods killed 53 and 84 people, respectively. Excessive heat, however, killed an average of 237 per year during the same time period, almost three times the number of deaths due to the much-feared tornadoes and hurricanes combined. Most of these heat-related deaths occurred in cities.

In the past fifty years, heat waves in the United States have been blamed for more than 400 deaths on four occasions. During a prolonged 1963 heat wave, more than 4,600 deaths occurred in the eastern United States. A stifling, lingering heat wave killed 1,700 people in the east and midwest regions of the country in 1980. In 1988, another east/midwest heat wave killed 454. In 1995, a heat wave claimed 716 lives in Chicago, Philadelphia, Milwaukee, and St. Louis, 600 of those in the Windy City alone. In the 1936 heat waves, more than 4,768 Americans died from heat and high humidity.

During these recent heat waves, most fatalities occurred in urban areas, where trapped heat brought unrelenting dangers to many, particularly the elderly and the poor, who could not afford air conditioning. Such figures worry bioclimatologists and public health officials. They fear that continued climate warming could increase heat-related fatalities, particularly in northern American and Canadian cities, where episodes of extreme heat are less frequent. Similar concerns have been raised in large cities outside North America, such as Rome and Paris.

A number of cities, including Toronto; Washington, D.C.; Philadelphia; and Rome, now use "Hot Weather Health Watch, Alert, and Warning" programs when weather forecasts indicate potential prolonged hot and humid conditions. The warning system is based on a *Weather Stress Index* (WSI) developed by Dr. Laurence Kalkstein and colleagues at the Center for Climatic Research at the University of Delaware. The WSI not only includes the absolute level of heat danger, but also incorporates an adjustment for the relative frequency of such conditions in the region. Kalkstein believes such warning systems could save as many as three hundred lives a year in the United States by issuing timely alerts.

Cooling the Cities?

Even without the specter of death associated with excessive urban heat, urban heat islands cost a city's residents money. The need for additional air conditioning can cost the typical household one hundred dollars or more per year, depending on location and local energy rates. In addition, power companies must maintain extra generating capacity that they may use only a few days each year, and this cost is always passed on to the consumer. For example, if hot days are too common in some urban areas, the number of rolling, and perhaps full, power blackouts may increase. Power problems could strike large regions of Canada and the United States if a prolonged heat wave were to settle over the eastern half of the continent, because approximately one-sixth of electricity consumption in the United States is used to cool buildings.

When the economic and health consequences of the summer urban heat island are tallied, the total is great enough to concern many urban governments and citizen groups. Several cities, such as Toronto and Chicago, have begun to look into ways of reducing the urban heat island's summer impact. One of the most promising initiatives is through the *greening* of the city, that is, by increasing the vegetation cover within urban limits. Adding greenery over an area as big as a city park or as small as an individual tree will help reduce the area covered by dark, heat-trapping surfaces and favorably move the heat balance toward cooler conditions. In addition to added vegetation, replacing dark, absorbing rooftops and pavement with lighter, more reflective materials can increase the city's *albedo* (solar reflectance), reducing the amount of sunlight absorbed by buildings and pavement. Rooftop gardens have also been proposed as a positive influence for heat island reduction as well as for air pollution reduction.

As cities heat under the summer sun, it's no wonder that urban residents flock to lake and sea shores or into the country. But, while such journeys are good for both mental and physical health, the heat generated by traffic and road systems adds to the discomfort of those left behind.

Forty-four

Bright Downpours

During a cloudburst, we are usually concerned with keeping dry or navigating our car safely through traffic. But have you ever watched the sky darken with a thunderstorm's approach and then be surprised at how bright the day becomes during the heaviest downpour?

Sunset and a small rainstorm across the Patuxent River, Maryland. (MARY HOLLINGER, NODC, NATIONAL OCEANIC AND ATMOSPHERIC ADMINISTRATION/U.S. DEPARTMENT OF COMMERCE)

I think we all expect that the pre-rain darkness of an approaching thunderstorm will deepen when the rain finally begins to fall and that, logically, it should become darker under a heavy rainfall. Very often, however, a sudden downpour from a towering cumulonimbus, sparking lightning and booming thunder, comes with an unexpected brightness.

During a heavy downpour, a large number of rapidly falling raindrops surrounds us. Each one can reflect, refract, and scatter light rays. We are, in effect, caught in the rain of thousands if not millions of tiny mirrors. As any interior decorator knows, you can brighten a room by putting mirrors on the surrounding walls or ceilings. Each light ray passing through the downpour undergoes multiple reflections, so the light reaching our eyes comes from all directions rather than just directly from the source.

The most important factors in producing the rainfall-brightening around us are the raindrop quantity and the size distribution. In a downpour, the number of large drops is generally much greater than in a light rainfall. Since a drop's surface area increases as the square of its radius, a drop twice as large has four times the potential reflecting surface area. As these liquid mirrors get bigger, we become more effectively surrounded by reflecting surfaces, and this too makes it appear brighter.

Forty-five

Twinkle, Twinkle, Little Star

I have often found September a perfect time for looking beyond my beloved atmosphere to the heavens above. The evenings are usually still warm enough to sit comfortably outdoors and gaze upward, yet sunset is earlier and full darkness comes quickly enough for me to enjoy the celestial show without staying up half the night.

The weather, of course, plays an important role in star and planet watching—an overcast night kills any chance of seeing beyond the cloud deck. Air pollution, combined with light pollution, also affects what or how much we see. Dust and other pollution particles in the air not only diminish the incoming starlight but can scatter urban light and render all but the brightest bodies invisible. But let's look upward this day from a location away from strong human influences. We'll also eliminate the clouds, although the right cloud-moon combination can bring many fascinating features to nighttime sky watching.

What is the first thing you usually notice about stars on these clear nights? I'll give you a hint from our youth: "Twinkle, twinkle, little star." Yes, stars twinkle. The "little" part may refer to the fact that the planets, which have much the same appearance as stars, do not twinkle.

The technical term for twinkling is *scintillation*, the rapid variation in apparent position, color, or brightness of a luminous object when viewed through a turbulent media, in this case, the atmosphere. Stars, as we know, are large masses of glowing gas similar to our sun, but they are so far away that they appear as bright pinpoints, even through telescopes. Starlight travels relatively straight and true through the vacuum of space across many light-years before reaching the top of Earth's atmosphere as a steady point of light (which is how the stars would appear to viewers aboard the International Space Station).

When starlight enters our relatively dense atmosphere, its rays are diverted from their direct path by changes in air density en route to the surface. This is called refraction. If the atmosphere was just a dense immovable coating around Earth, stars would appear slightly off their true location due to the refraction of the atmosphere. The atmosphere, however, is in fairly constant motion and becomes increasingly dense, though not uniformly so, as one moves closer to the surface. Light rays bend differently when they pass through cold and hot air regions, always bending toward colder air because it's more dense than warm air (assuming all other factors are equal). The horizontal and vertical motions of hot and cold air pockets cause light rays moving through the naturally turbulent atmosphere to change direction continuously.

The lower atmosphere is mottled with pockets of varying density caused by rising and falling air parcels and strong horizontal winds. When the local air density changes rapidly with time, a condition termed turbulent, the light ray's path also alters rapidly. This slight but perceptible refraction bends the path one way at one moment, a slightly different way the next. This constant but random shifting results in the star's image jiving and jiggling, fading in and out, even changing colors before our eyes. This stellar dancing is what we call *twinkling*. In addition to this constant jitter in the apparent position of a star, the turbulent air pockets also focus and de-focus starlight, making stars appear to randomly change brightness. Since the amount of refraction also depends on the light's wavelength, various colors in the ray may dominate at times, giving the twinkling star hints of color change. (This effect is not as apparent since our eyes are less color-sensitive in the dark.)

Often on a clear night with calm surface winds, wild twinkling of the stars indicates strong winds, such as the jet stream, high in the atmosphere. The greater the atmospheric turbulence, the greater the twinkle effect. Lots of twinkling stars indicate a very unstable, turbulent atmosphere.

Twinkling affects only those distant objects whose visual size is smaller than the refractive shifts caused by the atmospheric turbulence—that is why planets do not twinkle. Even though they appear as stars, they present a visually large disk compared to the level of turbulent refraction. Although their light still scintillates, the refractions of different light rays coming from across a planet's disk tend to cancel one another out, and the planet's light shines rather steadily. On rare occasions, turbulence may be strong enough to show some planetary twinkle, but usually only when the planet is near the horizon. A similar effect to twinkling, *shimmer*, can often be seen in distant surface light sources, such as individual city lights, when there is strong surface-layer turbulence.

Shimmer, over desert sands or other hot surfaces, is another form of scintillation.

Forty-six

Halos and Sundogs

As I sat on my balcony, scribbling away at my notes, I glanced up into the southern sky to see floes of icy-looking cirrostratus clouds drifting by. Beneath the solar orb, a brilliant spectrum of color interjected itself on the white and pale blue background. What I was seeing was an arc of a 46-degree halo, which did not completely encircle the sun. Shielding my eyes from the sun's direct glare, I could also see a 22-degree halo ringing the solar orb. (Common *halos* are named for the view angle formed between the solar center and the edge of the halo.)

Halos

In the halo family, the 22-degree halo, famed for its appearance in weather lore in "Ring around the sun or moon, rain will be a-fallin' soon," is the best known. The 46-degree halo forms outside the smaller halo, showing the same color sequence: red-orange on the inside, blue in the outer portion. But while the 22-degree halo generally forms a complete ring around the sun, the 46-degree halo is rarely seen as a complete encirclement, but is recognized by its arcing curvature. The cause of these and other halos is the refraction and spectral splitting of sunlight as it passes through ice crystals in the atmosphere. Today I spied the two halos; the ice crystals resided in the cirrus cloud deck above me. Within these clouds, the tumbling ice crystals had oriented themselves in the proper position for the sun's light to refract as its rays streamed toward my eye. The 46-degree halo formed from the same crystals as the 22-degree halo but arose from a different light path through them.

Ice crystals in the atmosphere are hexagonally shaped. The crystals forming most optical phenomena in the air are typically flat, hexagonal plate patterns, like microscopic stop signs or dinner

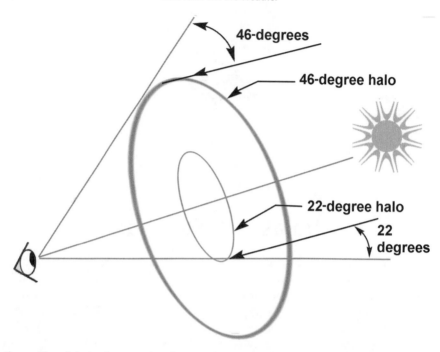

46-degrees

46-degree halo

22-degree halo

22 degrees

The position of the 22-degree and 46-degree halos around the sun as seen by the observer's eye.

plates, or hexagonal columns, shaped like common lead pencils. When plate-shaped ice crystals fall unimpaired, drag forces orient them so that their larger, flat surface parallels Earth. Column crystals tumble more when falling, often spinning like small helicopter blades. Ice crystals that form halos are usually columns, much longer than they are wide and very small (approximately 15 to 25 micrometres in diameter). The best viewing of the 22-degree halo occurs when a veil of cirrostratus lying across the sun contains an abundance of these uniform-sized hexagonal ice crystals. Light rays that enter one side face of the ice crystal and pass out another side form the 22-degree halo. Those that go in one side and exit one of the crystal end faces form the 46-degree halo. Since the 46-degree halo is rarer than its smaller partner, some investigators believe crystals must be "fatter" than normal (the *plate crystal*) to produce the larger variety halo. They argue that if the crystal is long and thin like a pencil, there is too little area on the ends of the crystal for the light to emerge through an end face rather than a side. The light that forms the 46-degree halo disperses over a wider stretch of sky and is generally fainter than the more common 22-degree form, which may account for its infrequent observation.

The 22-degree halo has long been considered a harbinger of rainy weather, although its appearance is not an infallible indicator. The reason for the link stems from the common sequence of cloud types seen with a warm front associated with a low-pressure system. In this situation, the warmer air behind the front overruns the colder air ahead and generally spreads into a wedge of clouds extending well ahead of the front. In the idealized sequence, wispy cirrus clouds appear first, then thicken into the cirrostratus veil, where our halos begin to appear. Upwind of the cirrostratus, we find lower alto-clouds, usually

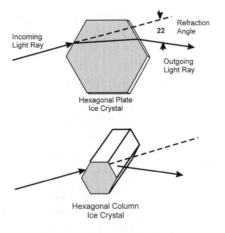

Light ray paths through hexagonal ice crystals which form 22-degree halos.

altostratus, which further lower to stratus, and finally the rainy nimbostratus in the vicinity of the front. Similar lore holds that if there is a halo round the moon, the number of stars within the halo indicates the number of days until the rain falls. This lore does not hold water (no pun intended . . . yes, it was!), although one could argue that the number of stars seen within the ring might indicate the air's moisture level and thus the distance to the front. If the cirrus layer was very thin, and likely farther out from the front, a greater number of stars might be visible compared to a very moist atmosphere with thick cirrus—likely closer to the approaching rainfall—that would block out most or all stars. However, the moon's brightness and the density of bright stars in the region of the heavens with the ring will also play a role in the number of stars seen, and this has no bearing on time until the next rainfall.

Sundogs

The same refraction process within ice crystals that forms halos produces a number of other optical gems in the air. Prominent among these are the "dog" family, the *sundogs* and *moondogs*, which flank the sun and moon, respectively. Both are loyal to their orb, usually sitting as pairs, one on each side along a horizontal line through the bright disk. Sundogs are most regularly seen close to their solar master during those low-sun months or hours when the sun is low in the sky, such as around sunset and sunrise.

Sundogs are technically called *solar parhelia* (*parhelia* meaning "with the sun") and appear as bright bursts of light formed when sunlight passes through ice crystals at the proper angle. Usually,

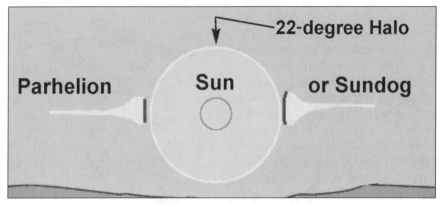

The location of sundogs relative to the solar disk.

cirrus clouds in front of the sun produce sundogs, but other ice clouds, such as ice fog and diamond dust, may also generate them. Sundogs are sometimes so brilliant that dazzled observers mistake them for a sun, giving them the alternate popular name of *mock suns*. They are often bright white but may show a partial spectrum of color, with a comet-like tail facing away from the sun.

Sundogs are the second most frequent halo phenomenon behind the 22-degree halo and often accompany that halo. The difference between sundog and halo formation is the orientation of the ice crystals through which sunlight passes before reaching our eyes. Halo formation requires a mixture of random ice crystal orientations in the sky. But if the sky has only horizontally oriented, flat ice crystals, we see a sundog. When plate-shaped ice crystals fall unimpaired, their larger, flat surface parallels the earth during descent like a large maple leaf drifting down from a tree. Sundogs emerge when sunlight passing through the ice plate's thin side faces is refracted. The more perfectly aligned the falling crystals are to the horizontal, the more compact the resulting sundog. Crystal misalignment from true horizontal will spread the sundog vertically—its angular height being approximately four times the maximum crystal angular tilt.

Sundogs frequently display a reddish tint on the side facing the sun and may sport bluish white tails which stretch horizontally away from it. The degree to which colors are visible depends on the amount of wobble in the falling ice crystals: the more wobble, the more color. The sundog's tail is formed by light passing through the crystal at angles other than the optimal deviation angle. Sundogs typically appear when the sun is low to the horizon, usually just prior to sunset or after sunrise, or during winter months at midlatitudes. If the sun is low (from the horizon to about 15 degrees above), each sundog is separated from the sun by 22 degrees (or about two handbreadths on extended arms) and lies on the circle of the 22-degree halo if one is present. Sundogs form tightest to the sun at the lowest solar altitudes, but they are never less than 22 degrees from it. As the sun climbs in the sky, the sundogs slowly move away from the 22-degree separation, although they remain on the line through the sun parallel with the horizon. When the sun has climbed to more than 45 degrees altitude, sundogs are fainter and noticeably off the 22-degree circle, and they vanish altogether above 61 degrees solar altitude.

Moondogs appear alongside the moon and are formed by lunar light passing through ice crystals, a similar process to the production of sundogs. Moondogs (or *paraselenae*) are less commonly seen because the moon can produce them only when bright and because they often appear late at night, when most of us are indoors or asleep.

Sundogs form along a faint 22-degree halo at sunset in Davenport, Iowa.
(US NATIONAL WEATHER SERVICE FORECAST OFFICE, DAVENPORT, IOWA)

Forty-seven

The Fog Comes In . . . or Down . . . or Up

As I was heading home around the witching hour recently, I noted that the clear sky above was hazier as I looked toward the horizon. Dew was beginning to form on the windshield, so I knew the surface air was fairly close to saturated. I suspected fog would form around the area overnight. The sun's tenure in the day declines steadily now as we approach the equinox, when it relinquishes majority control to the night. Fogs form more frequently during this time of year because, with the longer night hours, Earth's surface can cool for a longer time during clear and calm nights. When the cooling reaches the dew point, fog forms across the land.

The specific type of fog I expected this night is termed *radiation fog* and is quite common across Canada and the United States, particularly in autumn. Fog, however, comes in many flavors, depending on the prevailing large-scale weather regime, location, and topography. Radiation fog's cousins, *frontal fog* and *advection fog*, also find the year's last quarter conducive for a visit.

What Is Fog?

Although it may not be recognized as such, fog is actually a cloud formed or lying on the ground—even those patchy fogs that fill low spots or hollows in the terrain are really cloud fragments. Fog formation depends very strongly on two characteristics of the air mass in which it forms. Foremost is the humidity; if the air mass is very dry, the likelihood of producing fog is slight. To form fog, the air mass must be brought to a saturation condition with respect to its humidity and temperature. Thus, more and more water vapor can be added until the air becomes saturated, or the air can cool until its temperature reaches saturation point. At saturation, water vapor begins to form liquid water droplets, which comprise the fog we see.

Fog is defined—for weather-observing purposes—as an obscurity in the atmosphere's surface layers caused by a suspension of water droplets. It is defined by international agreement as being associated with visibility less than one kilometre (0.62 mi.). Often, we see foggy patches that dot the landscape or hang over one area. This is also fog, even though it does not fit the visibility observational requirement (though it likely would if larger in extent) and would likely be reported as "patchy fog."

Droplets that form fog and clouds are very small, only from 1 to 20 micrometres in diameter, and thus fall to the ground very, very slowly. This allows fog and clouds to seemingly hang suspended in the air. If the air warms slightly or mixes with drier air, the liquid water can rapidly return to its vapor state. This is why fogs often quickly dissipate with the morning sun or an increase in the wind. As a result of this delicate balance, cloud/fog edges are tenuous and ever-changing.

The condensation of water vapor in an air mass can be produced by several mechanisms, the most common of which are:

- the ascent and resultant cooling of an air parcel to its dew point
- radiative heat loss from the air parcel, which cools it to the dew point
- the mixing of two parcels of slightly unsaturated air initially having different temperatures, usually a cool, slightly drier air mass mixing with a moist, warmer one
- the descent of cold air, through the pull of gravity, into a valley or low terrain basin where surface moisture is present (such as a lake, pond, or marsh)

Radiation fog and frontal fog are perhaps the most common fog types in areas away from large water bodies. In hilly or mountainous terrain, one might experience *upslope fog* or *valley fog*. Over and along the shores of large water bodies, fogs may be termed *advection fogs*, *sea fogs*, or *steam fogs*, and they form by one or more of the above processes. Depending on how a particular fog forms, it can rise or fall or roll in, or even just envelop.

Radiation Fog

Radiation fog generally forms when the air near the surface cools to its saturation temperature by radiational cooling at night, after the sun has set (or is close to the horizon in circumstances when the daylight temperature is very close to saturation temperature). A typical scenario for radiation fog might look like this: We start with a

Radiation fog envelopes a neighborhood street in Ann Arbor, Michigan.
(KEITH C. HEIDORN)

clear evening with moderate or high humidity and light winds. As the sun's energy diminishes at dusk, the ground surface radiates its heat away faster than it can gain it from other sources, and its temperature drops. The cool surface then chills the air in contact with it. Throughout the night, the surface and overlying air continue to cool (unless warmer air moves in); the degree of cooling depends on several factors, including cloud cover, wind speed, and number of hours of darkness.

If the air reaches its condensation, or dew point, temperature during that nocturnal cooling, fog will form. A wet surface—moist soil or pools of standing water—significantly increases the chance of radiation fog formation, so radiation fog potential is high after a rainfall, particularly if the rain is followed by a cold front that clears the skies and lowers the air temperature. Radiation fog is not common over water surfaces since the cooling of the water surface by radiation is much slower during the night than the cooling occurring on the land. However, air cooled by radiation loss may flow down terrain to settle and form a fog over a pond or lake.

The radiation fog layer varies in depth and horizontal extent, from shallow scattered patches formed in surface depressions to a general blanket as deep as 300 metres (984 ft.). Visibility within the fog drops to less than 1 kilometre (0.6 mi.) as the density of water droplets in the air increases. With further cooling and more condensation, the fog layer becomes optically denser, often reducing visibility to the proverbial "can't see one's hand in front of one's face" state. Often, and particularly in autumn, when long nights are conducive to deep cooling, we see patchy fog in the countryside form in low-lying areas. These are preferred fog locations because cold air, being denser than warmer air, flows like water toward the terrain's lowest point. Therefore, if there is any slope to the land, such as water drainage channels or hillsides, the cooler air will flow into the lowest elevation points.

Nocturnal radiation fogs often bypass cities, while rather thick fog forms in surrounding rural and suburban zones, because the city produces and holds heat much better than more open, vegetated suburbs and does not cool as deeply overnight—this is the urban heat island phenomenon mentioned on page 183. Air over a city is also often slightly drier than that over its environs.

Radiation fog dissipates when the air mass is reheated, or the wind increases speed and mixes the moist, foggy air with warmer or drier air around it. Where fog is thin, morning solar radiation can penetrate to the ground and heat the underlying surface, thus evaporating the fog from below, a process that can be rapid in summer and excruciatingly slow in winter.

Frontal Fog

Frontal or precipitation fog generally occurs when rain falling from warm air aloft evaporates at or near the surface under light wind conditions. As it falls through the colder air, the evaporating precipitation increases the surface air's moisture burden until condensation is achieved. Such fogs are most common in the vicinity of slow-moving warm or stationary fronts, but they can form at cold fronts as well. However, cold fronts generally move and mix too quickly to allow the condition to persist for long. Precipitation fog can also occur under other conditions, such as beneath an area where rain falls from air driven upslope over terrain. It may also form briefly in areas where hot surfaces are quenched by showers. In this situation the hot surface forces rapid evaporation of the rain hitting it, and the vapor mixes with the cool air surrounding the falling rain to become oversaturated.

Upslope and Valley Fogs

Upslope and valley fogs are two special cases, in my mind, of advection fogs particular to hilly terrain. They form when air moving in hilly or mountainous terrain cools to condensation. When air moves over a terrain obstacle, it cools to some degree as it rises, the degree of cooling depending on the height of rise. During that cooling, if the air temperature falls below the dew point, the resulting condensation will form a cloud. If that cloud hugs the ground, it becomes upslope fog at the terrain surface. For example, an air mass moving over a water body may gain moisture until it is very near its saturation level. While still over the water, it may not cool enough to reach condensation, but when forced over the shore, the rise from water level may cause enough cooling to form a fog. When you see clouds hugging a mountain summit, you can be sure it is foggy on the mountain they embrace. In areas like the raincoast of the Pacific Northwest, upslope movement of moist Pacific air can cause extensive fog at higher elevations, which disappears when the air descends on the lee of the ranges.

Fog and haze form over the Georgia basin on a winter's day when warm waters are overlain by cold air. (Keith C. Heidorn)

Valley fogs form when the air near the terrain heights cools—usually by radiation at night—and descends through its greater density into the surrounding valleys. Pooled in the valleys and becoming progressively colder, the cold air may condense the water vapor present into a fog that fills the valleys to a significant depth. Satellite photographs often show the dramatic regionality of valley fog, bright fingers of fog lying between mountain ridges.

Advection Fog

Advection fogs are formed when moist air moves over a cooler surface or cool air rises over a warmer, moist surface, and the air mass reaches saturation as a result. Most often this occurs when a moist air mass moves over a large, cold body of water or extensive snow or ice cover, where the temperature is below the dew point of the advecting (advancing) air mass, and its lowest reaches are cooled to condensation. The formation of advection fogs is enhanced when the distance (*fetch*) over which the advecting air moves is large. A low wind speed heightens the likelihood because the air remains in contact with the surface long enough to cool the air layer sufficiently. Since the weather situation that forms them can last a day or more, advection fogs are often persistent. Usually, either a frontal passage and change of air mass or a major change in wind direction is needed for dissipation of advection fog to take place.

Evaporation advection fogs occur when a cool or cold air mass crosses a relatively warm water body and has water vapor evaporate into it. Often, such fogs are caused by cold air outbreaks moving off land to cross large water bodies such as the sea or large lakes. These advection fogs are often referred to as *sea fogs* and are common offshore in winter. Sea fog is sometimes called *sea smoke* or *steam fog* when it rises from the water surface in plumes of moist air that resemble smoke.

Whichever mechanism gives rise to fog, local and regional features can influence its extent and thickness. When light and patchy, fogs can add a beauty to the landscape. When thick and widespread, they can make travel hazardous and even accentuate poor air quality conditions. No wonder "fog" has many non-weather meanings in English.

FALL

DUST DEVILS • SNOW PILLARS • BLIZZARDS • CLIPPERS • SNOWFLAKES • CREPUSCULAR RAYS • MOONBOWS • HALOS • SUNDOGS • RAINBOWS • LIGHTNING • THUNDERSTORMS

Forty-eight

22 September—The Autumnal Equinox

Our journey through September eventually brings us to 22 September, the autumnal equinox, the day when the sun again shines overhead at the equator. For many, this event signifies the start of autumn or fall, the last full quarter of the calendar year.

Several years ago, I was admonished by an Australian colleague for calling the season "fall." "In many parts of the world," she said, "leaves do not fall in this season." However, my research indicates the term was not applied to the season because of the fall of leaves, but referred to the sun falling below the equator after this date (from a northern hemisphere perspective, anyway). The phrases "spring up" and "fall down" are reminiscent of that daylight saving time mnemonic "Spring ahead, fall back."

If you follow the solar seasons, this is the mid-autumn day, the time when the daily rates of solar waning (which leads to "solar *whining*" on my part) and day length change are at their peak. If nothing else, to solar people, the time signifies the six-month stretch when night hours are in control. True, the actual first day when night exceeds day is a few days off due to atmospheric refraction effects, but the equinox still holds its mystique.

Others have already declared summer over with the passing of Labor Day, though I have never seen the media describe these intervening days between the "official" end of summer and the "official" start of autumn with any "official" name. We are hard pressed to place a true start to the new season each year; however, I'm becoming more convinced that the change is not abrupt but slow, almost imperceptible, so that one day you realize autumn has arrived. (I see this similar process in my aging. I now admit to being in my autumn years, though I still await an Indian summer reprieve.) The coming of the equinox may signify a changing of the seasonal character, repeating hints of cold weather more frequently or bringing the first dusting of snow, or perhaps continuing the last of summer's heat, depending on where in North America one lives, but it is not generally an on-off switch.

Just past mid-month, September eventually surrenders summer to the agents of autumn and winter. First frost will soon raid southward, driven by arctic winds or pouncing on a clear, still night. Snows will whiten the Canadian prairies and cross the international border into Wyoming and Montana, sugarcoating mountain summits as far south as Colorado and at times dusting the Great Lakes and St. Lawrence Valley countryside. In some years, Jack Frost has already come to town when the equinox arrives, and with his deadly touch ends the life of many plant and some animal species. In many others, the calendar slides into October before we encounter the finale. In either case, Jack's visit prepares us for the encore of Indian summer.

Forty-nine

Painting Fall's Colors

You can see it distinctly on the weather map or satellite picture pushing southward over eastern North America: a large high-pressure mass of arctic air that brings the first hard frosts to the Great Lakes, New England, and perhaps into the Ohio Valley. Such autumnal air masses, characterized by clear skies and temperatures that are often pleasantly warm during the day and chilling at night, carry the first widespread hints of autumn's brilliant colors.

Imagine looking across a field still strewn with stalks of now-harvested corn. There in the distance stands a small woodlot of mixed deciduous trees. Each of the species represented in the community wears its full blaze of autumnal hue. The scarlet of red maple and the orange-red of sugar maple mix with the gold of ash and the yellow of birch. The perpetual green of the cedars intertwines with their deciduous neighbors, completing the kaleidoscope of color. What magic transforms the uniform green of summer to this annual spectrum of color? Contrary to popular belief, Jack Frost's brush does not cause the outburst of autumn color. The brilliant vistas of autumn are the product of subtle changes in internal leaf chemistry as deciduous trees prepare for winter dormancy, the signal initiated most often by diminishing daylight hours. (A dramatic example of this is the increased time trees retain their green leaves when surrounded by bright street lamps.)

Through a process known as *abcission*, leaves fall from the trees, and it is the preliminary phases of this process that change leaf chemistry. Abcission begins in early autumn. Responding to the shortening days and declining intensity of sunlight, a leaf's veins, which carry fluids between the branch and leaf, gradually close off as a layer of cells forms at the base of each leaf stem. When the separation layer is completely formed, the leaf is ready to fall. The sealed veins trap sugars in the leaf, promoting the production of pigments known as *anthocyanins*, the chemicals responsible for autumn's red colors.

During spring and summer, as emerging leaves unfurl the great sugar factory of photosynthesis, trees produce chlorophyll, which gives them their distinct green hue. In autumn, as night gains dominance over daylight, production of new chlorophyll wanes and then ceases, while the decomposition of older chlorophyll continues; eventually, no more remains. Without other pigments, the leaves would pale and soon turn colorless.

However, as the chlorophyll disappears, three pigments gain prominence, their colors becoming the colors of autumn: *carotenoids*, *anthocyanins*, and *tannins*. Carotenoids are responsible for the yellow hues of the season as well as the coloration of carrots, corn, daffodils, rutabagas, buttercups, and bananas. These pigments assist chlorophyll in absorbing sunlight and are present all summer, but become visible only when chlorophyll is no longer dominant in the leaf. Anthocyanins are responsible for the scarlet, blue, and purple hues in nature. They comprise the characteristic colors of grapes, cranberries, blueberries, red apples, cherries, strawberries, and plums. Unlike carotenoids, anthocyanins must be produced within the leaf after the chlorophyll has declined and are formed by the breakdown of excess sugars under bright light. Bright autumn days with cool nights favor their formation, thus the connection between brilliant autumn hues and Indian summer weather. Tannins are the third prominent pigment, contributing brown tones to leaves, especially those of the oak and elm. After the carotenoids and anthocyanins have decomposed, the tannins alone remain, giving the dead leaves their characteristic brown color.

Vivid red, purple, and crimson colorations from anthocyanins are stimulated by abundant bright sunlight and high sugar content, while duller colors result from the effects of cloudy skies and shading. Bright sunlight is so essential for the production of the anthocyanin pigment in leaves that if a portion of the leaf is prevented from receiving sunlight before it turns red, that portion will turn yellow while the exposed part will turn red. Since carotenoids are always present in leaves, the yellow and gold colors remain fairly constant from year to year.

The extent and brilliance of autumn colors in any particular year depend on weather conditions that occur before and during the time the chlorophyll in leaves is dwindling. Temperature and soil moisture are the main influences affecting color intensity. Temperatures much below freezing will kill the leaf cells before they can produce these pigments, while an unusually warm spell may lower the intensity of autumn colors. Like the weather, soil moisture varies greatly from year to year. A severe late spring or summer drought can delay the onset of autumn color by a few weeks. A warm wet spring, favorable summer weather, and warm, sunny fall days with cool nights generally produce the most brilliant autumn colors. Soil moisture may also vary with the topography, affecting the coloration within a small area to differing degrees.

The timing of the color change also varies by species. Oaks, for example, change their colors long after other species have shed their leaves. Differences in the timing among species seem to be genetic and driven by day length; so, a particular species at the same latitude will show the color changes in the cool temperatures of high mountain elevations at about the same time as those in warmer lowlands. The particular combination of weather, species, and other environmental factors makes each autumn unique, and each forest or woodlot special. Colors may vary in shade and intensity on opposite sides of a valley or even on the same tree. Thus, the infinite variety of autumn colors is born.

Fifty

Riding the Sunbeams: Crepuscular Rays

Here on Vancouver Island, the onset of autumn signals the end of frequent clear, dry days and the start of the rainy season. Clouds will thicken and lower, becoming ever darker as their water burden increases, and soon rain will be commonplace and sunlight too rare. One ray of sky glory common during these months is the brilliant display of visible solar rays, often tinted red or orange in the twilight hours. They frequently break through during those last (or first) minutes of daylight and early minutes of twilight, vividly illuminating the base of an altocumulus or stratocumulus deck.

Crepuscular rays fan out from low sun behind a line of cumulus clouds. (KEITH C. HEIDORN)

Solar rays are something every child learns to add to drawings of the sun. We name them *sunbeams*, the *Ropes of Maui*, the *Rays of Buddha, Jacob's Ladder*, and, more technically, *crepuscular rays*. To see them, we need only the sun, something to cast a shadow, plus a little dust or other particles in the air to make the rays visible. Sunbeams appear as light or dark shafts that seem to radiate from the sun. Although they can be seen anytime the sun is in the sky or just below the horizon, sunbeams are most commonly seen, and often their most beautiful, around sunrise and sunset. Indeed, crepuscular means "related to twilight."

To see sunbeams, a shadow must be cast. In the skyscape, clouds are the usual shadow-makers, although an irregular mountain range may also paint sunbeams across the sky when the sun is low behind them. If you walk or drive through a forest or a dense woodlot, sunbeams may appear filtering through the trees. Windows or skylights invite sunbeams indoors, and in some cases, building designs have included sunbeams as part of the ambience. Whatever the caster of shadows, its critical role is to break up the sunlit sky into regions of light and dark.

Solar rays outside the direct beam, however, are not visible on their own. That requires the assistance of small particles—atmospheric gases, dust, water droplets, snow, or ice crystals—to scatter or reflect light to our eyes. For example, dust particles can often be seen dancing in a sunbeam shining through a window because they reflect the light toward our eyes. Generally, sunbeams appear brightest when a dark background, such as a black cloud, provides greatest contrast. A dark background may also bring out very vivid colors in crepuscular rays when sunlight in the red-yellow portion of the visible solar spectrum crosses dark clouds. A beam's contrast also depends on the density of scattering particles, the scattering angle of the sun, and the line-of-sight distance through the light ray. The highest degree of contrast occurs when the rays are viewed toward or away from the sun, rather than when viewed at right angles to the beam, such as when they pass overhead.

The pattern recognition functions of our brain may also cause us to perceive "dark rays." These are usually seen when scattered clouds intercepting portions of the sun's radiance cause us to interpret their shadows as "rays" rather than the bright sky patches between the shadows. Another illusion arising from sunbeams is the perception that they converge on the solar disk. In fact, light rays emanating from the sun are parallel when they strike Earth. That sunbeams appear to converge on the sun is due to *perspective*—the apparent convergence of parallel lines at some distant vanishing point. This is the same visual effect that causes railroad tracks to appear to converge in the far distance.

The most beautiful and striking visualizations of sunbeams are the crepuscular rays seen when the sun is near the horizon. Crepuscular rays are usually red, orange, and yellow because blue light is effectively scattered out by air molecules and very small particles in the sky. Although we see crepuscular rays diverge from the sun toward us, if we turn around, we may see them converging toward the eastern sky. These portions of the rays that converge on the antisolar point of the sky are called *anti-crepuscular rays*, which appear as a pastel-pinkish glow against the darkening blue sky.

A Delicate Balance

Among autumn weather's prime features are the ever more frequent visits of the cold air masses that bring frosty weather across the continent. These nippy air masses have been building in the northern regions, in corners tucked away from the Prevailing Westerlies, ever since the sun's strength diminished to the point it could no longer provide sufficient energy to overcome nighttime cooling. While Earth receives warming solar energy during daylight hours, it radiates away heat energy both day and night at a rate dependent on its temperature, surface characteristics, and local moisture conditions. Among the classic features of autumn, fog, dew, and frost may form as a consequence of nocturnal thermal radiation exchanges near the surface. These exchanges lower surface and air temperatures until atmospheric water vapor changes to a liquid or solid state.

The *radiation balance* is a basic concept for many weather and climate topics, including the high-profile concern over so-called *greenhouse gases*. Close to the planet's surface, the radiation balance plays a major role in determining the weather conditions we feel. In autumn, the radiation balance takes on a quickly changing nature as summer wanes and winter emerges.

First, let's step back from Earth, way back, about 150 million kilometres (93 million mi.), to our own star, the sun. The sun produces a tremendous amount of energy in its solar furnaces, which it emits as shortwave radiation across the universe. An extremely small fraction of overall solar output falls on the sunlit side of Earth, some of which immediately reflects back into outer space. The amount reflected, known as the albedo, is about 36 percent for the whole planet, but can be above 90 percent for bright surface areas such as snow, ice, and overlying clouds, and 10 percent or less for dark rocks and forests. The remaining fraction of the *incoming solar radiation*, *insolation*, heats Earth through absorption by atmospheric gases, water, rock, soil, vegetation, and building materials.

If Earth were not able to lose that energy through counter-radiation, the planet would long ago have become a molten ball and eventually vaporized and dispersed into the farther reaches of the solar system. Fortunately, the physical nature of all matter provides a remedy. Everything, from galaxies and gas giants to molecules and atoms, radiates energy at a rate dependent on its physical nature (such as density or molecular and atomic structure) and its temperature. Everything in the universe sends its heat energy out to its surroundings and receives energy back from every object in its view. When more energy is gained than lost, the body warms. When more energy is lost than gained, the body cools. When there is a balance between gain and loss, the body can maintain a constant temperature.

This basic concept applies best on a system more or less in radiative equilibrium, or when viewed over long time scales, but on our home planet, short-term, fluctuating radiant energy emissions are the norm. We see this every day. As the sun—a huge source of radiant energy—moves across the sky toward setting, our little local patch of planet receives varying amounts of insolation. If we consider daily and hourly differences in radiation acting over the whole planet, and also account for seasonal cycles and cloud-cover variations, it becomes obvious that the insolation energy received can be quite variable from one location to the next on a day-to-day basis.

The unequal quantity of insolation received across the globe produces well-known temperature variations: typically hotter during the day and in equatorial regions, colder at night and at the poles. The Earth system tries, albeit never totally successfully, to redistribute that heat energy from warm to cold regions so that the planet can establish a uniform temperature. Two fluids are called upon to transport that heat: water in the oceans and gases in the atmosphere. In the atmosphere, the energy redistribution process (generally expressed as heat transport) is known as the *weather*.

Let's concentrate on the radiation balance of a small parcel of the Earth system surrounding us. During daylight hours, the insolation entering our parcel of air, minus the reflected part, is absorbed by atmospheric gases and particles and surface materials (rock, water, soil, vegetation). *Thermal* (heat or infrared) *radiation* (emitted by the planet and its elements as longwave radiation) is also exchanged among the atmospheric constituents and surface elements. Since the rate of energy gained and lost varies among these elements, portions of the system may store some energy as internal heat for a time. With the sun in the sky providing a strong energy input, there is usually a net gain of energy in the atmosphere during most of the day, particularly at the surface, where air molecules are warmed

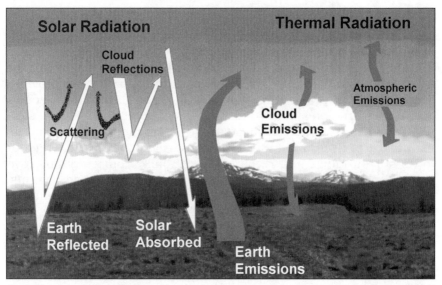

Daylight solar radiation balance and terrestrial radiation balance. Arrows pointing to the surface indicate incoming radiation; arrows pointing upward indicate escaping radiation.

through heat conduction and convection from the solid or liquid surface beneath.

At night, we have a very different picture. The sun's direct energy input is cut off, and the strongest energy input to the surface ceases. Earth's surface, however, continues to radiate heat away, and thus cools. How much energy is lost depends on the length of the night and the nature (including the temperature) of the atmospheric and surface elements. Rock and snow, for example, radiate heat differently, with snow losing its heat much faster. Water in the atmosphere (clouds and vapor) absorbs thermal radiation well and re-radiates a portion back toward the surface. Thus, the surface generally regains more of its lost heat radiation on humid or cloudy nights, and the air in contact with the surface cools more slowly. Maximum radiational cooling of the ground and the surface air is observed on long, clear nights with low humidity. Cooling under these conditions is often rapid at first, then continues slowly throughout the night until an hour or so after dawn.

Now it may seem logical to assume that the radiation balance turns from positive to negative at sunset and returns to positive at dawn, but this is not the standard case. For an hour or so before sunset (and after sunrise), depending on latitude and season, the insolation received at the surface is too weak to compensate for the thermal radiation heat loss. As a result, on mornings when the radiation process is controlling the local temperature cycle—as opposed to a change in air mass—the minimum temperature is often recorded in the hour or so after dawn.

In autumn, with darkness exceeding daylight, more time is available for the surface temperature to drop below the dew or frost points as a result of nocturnal radiation loss. In a similar process whereby the daily radiation balance turns negative, the seasonal radiation balance also makes such a turn over the course of the year. In solar summer, the balance is positive, in solar winter, negative. But in the transitional seasons of autumn and spring, there are delicate periods of balance. Consequently, the number of occasions with heavy dew, light ground frost, and patchy fog increases as we move through the autumnal season.

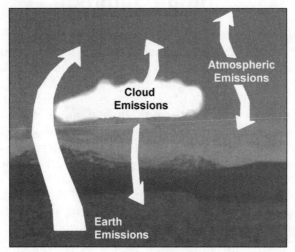

Nighttime terrestrial radiation balance. Arrows pointing to the surface indicate incoming radiation; arrows pointing upward indicate escaping radiation.

Fifty-two

Indian Summer

While I lived in the Great Lakes basin, one of my most anticipated annual events was Indian summer. Indian summer brings the region a stretch of several days with warm afternoons, often with a strong touch of humidity, and evenings mild enough to keep the windows open well after dark. Adding to this pleasant weather, the bright visage of autumn crimson and gold foliage against rich blue overhead skies treated all my senses to a vast panorama of sensuality. Should Indian summer coincide with a full or nearly full moon, the pleasant evening is ushered in with a soft yellow- or orange-hued lunar orb rising large on the eastern horizon.

Indian summer is not, however, a strictly defined meteorological entity and is as much an emotional event as a scientific one. Sample the population in North America for a definition, and you'll likely get a plethora of thoughts. I am hard-pressed, for example, after a decade on the West Coast, to define Indian summer here because hard frosts are rare in autumn. If I could, surely it would be quite different from one derived from my four-plus decades living among the Great Lakes. I have no problems, however, in defining Indian summer from my experiences there.

The Weather of Indian Summer

The weather scenario for Indian summer unfolds with the relaxing of winter's first major bite on the countryside. Frost has been hard and snowflakes may have danced in bitter northerly winds. But, as the cold air mass flows off the Atlantic coast, it is replaced by an extension of the subtropical Bermuda High, and a warm, humid southerly airflow becomes established. What makes this situation different from the wedge of warm tropical air found in an approaching low-pressure system is that the Indian summer regime locks in for days.

The warmth of the Indian summer air mass is deep—extending high into the atmosphere—and large, often covering the region from the upper Great Lakes south to the Gulf of Mexico and along the Atlantic coast. The air mass's presence alters the flow of the polar jet stream, pushing it farther north and locking in Indian summer for a stretch. This large, warm air mass, characterized by clear skies and light winds, stagnates over an extensive area and may remain unmoved for five to ten days, perhaps as long as two weeks or more.

Within the air mass, skies are quite clear, often colored deep blue, with a hazy fringe around the horizon. In the air mass's upper reaches, the air sinks and in doing so warms to form a *subsidence inversion* above the surface, which strongly inhibits cloud formation. Winds are light and variable, often becoming calm for long periods during the extended autumnal night. After sunset, rapid surface cooling will often form a deep surface inversion layer, and fog patches frequently arise within it.

The downside of the Indian summer air mass is that it is conducive to the accumulation of air contaminants. Within the stagnant air mass, light winds, nocturnal surface inversions, and an air mass subsidence inversion may combine to accumulate air pollutants, giving a hazy look to the sky around the horizon and, in many places, producing very poor air quality after several days.

Eventually, a heavier cold air mass pushing out of Arctic regions breaks the domination of the Indian summer air mass. Arctic air steers the jet stream and its attendant storm systems back south, and "normal" autumn weather returns with a new wave of frosty nights.

No criteria have been formally established for what constitutes Indian summer and, of course, any generally accepted criteria may vary from locale to locale. The broadest definition of Indian summer conditions includes:

- previous hard frost
- mostly clear skies (or perhaps local fog at night and early morning)
- no precipitation
- light winds and generally calm nights
- daytime maximum temperatures between 18 and 22°C (64 and 72°F), perhaps higher
- nighttime low temperatures between 2 and 9°C (36 and 48°F)
- event lasts for at least three days

The high and low temperatures are ballpark figures and vary as one moves north or south, but the Indian summer period is usually considered frost free.

A Weather Singularity

Indian summer is an example of a *weather singularity*; that is, a discernable weather event that recurs around a specific calendar date each year. Winter's January Thaw and spring's *Dogwood Winter* are other well-recognized weather singularities in North America. The latter half of October has the most frequent occurrences of Indian summer in eastern North America; however, it can arrive anytime between late September and mid-December. Not all years will have an Indian summer, and some may have two or three such periods.

Some Indian Summer Lore

The origin of the term "Indian summer" is uncertain. One common explanation suggests that Native Americans recognized the pattern, which they attributed to the good graces of the god of the southwest. When the eastern tribes described it to the first European settlers of Atlantic North America, the event became known as the "Indian's Summer." Another explanation attributes the name to the belief that the haziness of Indian summer days was caused by prairie fires deliberately set by midwestern tribes.

Weather singularities similar to Indian summer are known in Europe and take on a variety of names, such as *Old Wife's Summer* and *Second Summer*. In Poland, the period is called *God's Gift to Poland*, while the English call it *All Hallow Summer* or the "Summer

of" the saint whose day falls closest to the autumnal period when Indian summer weather occurs. Central Europeans often refer to similar weather conditions as the *halcyon days*, harking back to Greek mythology.

Indian summer weather paints a peaceful autumn scene at Bronte Creek, Ontario.
(KEITH C. HEIDORN)

Fifty-three

Heavenly Shades of Twilight Time

With the onset of autumn, shortening daylight hours allow many of us to view both sunrise and sunset. Often, October skies are clearer than they will be during the winter months. While you might think clear skies are not conducive to sky watching, there are subtle features in the clear sunset and sunrise skies that are worth looking for.

Before we look at the atmospheric effects, I want to define *twilight* because it has strict astronomical and legal definitions that are often used in city bylaws. Astronomers and timekeepers define three twilight periods: *civil twilight, nautical twilight,* and *astronomical twilight.* When evening twilight ends, the night begins, and it becomes *officially* dark until the beginning of morning twilight. The duration of each twilight period depends on latitude, elevation, and season.

Civil twilight is the period from sunset when the solar disk has just left the horizon until its center is 6 degrees below the horizon. If the sky is clear, normal outdoor activities during civil twilight can be continued, including the ability to read normal type without artificial light. For the sunrise period, the sequence is, naturally, reversed, and civil twilight begins when the sun is 6 degrees below the horizon until its upper limb just touches the horizon. *First light* marks the beginning of morning civil twilight, and *last light*, the end of evening civil twilight.

Nautical twilight occurs when the sun is between 6 and 12 degrees below the horizon. The brighter stars become visible during this period, providing good conditions for position fixing using manual navigation instruments.

Astronomical twilight fills the time interval when the sun is between 12 and 18 degrees below the horizon. During this time, only the gross outlines of objects can be discerned. When the sun is below 18 degrees, it is officially dark or "night." For regions in higher latitudes during summer, the sun may set, but no period of official dark occurs as the twilight periods of evening and morning merge.

The Twilight Sky

Under clear skies, the twilight period can provide subtle sky beauty with its softly changing colors, particularly the sky regions known as the *twilight* and *anti-twilight* arches. When the sun is on the horizon, the sky surrounding the solar disk takes on an orange-yellow glow. The colors in the red-yellow end of the spectrum dominate because the intervening air has scattered all the blue wavelengths from the sunlight before it reaches our eyes.

As the orange-yellow sun sets, the sky above it glows a pale yellow with yellow-orange patches to either side of the solar disk and a topping blue-white arch—the twilight arch. As twilight progresses, this twilight arch becomes pink with yellow and orange beneath. The twilight arch forms when sunlight is scattered by the atmosphere and usually begins encircling the sun like an aureole. When the sun drops below the horizon, the red wave bands are scattered toward us, often producing a coppery or blood-red twilight arch. The final glimmers of sunlight on the horizon may be tinted greenish yellow. On rare occasions, and usually when sunset or sunrise is viewed over water and the air is free from any form of haze, a quick green flash is seen on the top of the sun's disk just before it totally disappears from view. As twilight continues, the twilight arch slowly flattens, and the sky above darkens from blue gray to a deep blue merging into the darkness of night.

The Anti-Twilight Arch

If you turn your back to the setting sun, you see an opposing effect: the anti-twilight arch and the rising *Earth shadow*, the darkest region in the eastern sky. Earth's shadow, cast onto the atmosphere by the setting sun, appears as a large blue-gray arc, sometimes tinged with violet. Its brightness is caused by the backscatter of some of the sunlight by the relatively thick lower atmosphere through which the rays pass and is most pronounced when the eastern air is hazy. The Earth shadow seems to rise rapidly. Soon its edge becomes less distinct and disappears when the shadow limb reaches about 10 to 15 degrees above the eastern horizon.

The Earth shadow's upper edge is often bordered by a pinkish ribbon, the anti-twilight arch, above which a faint purple or yellow tint is sometimes discernible. It begins as a thin line stretching nearly 180 degrees in the sky opposite the sun and rises as the sun sets. The anti-twilight arch divides Earth's shadow from that part of the high sky still lit by direct sunlight. It is highest at the antisolar point (the height above the horizon equal to the sun's angle below the horizon) and curves down toward the horizon. Initially, the arch boundary is fairly sharp and edged with a reddish band, the *counterglow*, which becomes diffuse as it rises. As the sun drops below the horizon, the anti-twilight arch rises and becomes less distinct, finally blending smoothly into the dark sky of night.

A Jet Stream Runs Through It

High above Earth's surface, rivers of air rush their way around the globe in a high-speed current that often spawns side eddies that we on the ground call *cyclones*. Knowledge of the position of the most rapid flow, known as the *jet stream* because of its speed, is vital to accurate general weather forecasts and severe storm predictions. Knowledge of the location of the jet stream is also important to the airlines as its force can hinder or aid an aircraft's speed. In North America, the jet stream is a common topic for discussion in media weather reports, but they often give us the impression there is just one such stream. In fact, both hemispheres have two major jet streams (which may sometimes branch): the *polar jet stream* and the *subtropical jet stream*. On occasion, a fast-moving current in the lower atmosphere may also form, and this is known as a *low-level jet*. When mentioned in North American weather forecasts and discussions, "jet stream" usually mean the polar jet stream.

The jet stream is generally defined to be a current of fast-flowing air at high altitudes, somewhere 8 to 15 kilometres (5–9 mi.) above Earth's surface. Jet stream wind speeds blow, by definition, usually in excess of 90 km/h (56 mph) and can reach nearly 500 km/h (312 mph). They are quite variable in their dimensions, but a typical jet stream is hundreds to thousands of kilometres in length, about 160 to 500 kilometres (99–312 mi.) wide and about 1 kilometre (0.62 mi.) deep. The jet winds usually have a west to east direction, though they may loop with large north-south deflections.

The subtropical jet stream generally forms during the hemisphere's cold season. It is initiated by warm air flowing away from the tropical regions. This poleward airflow, however, is steered off its north-south course by the Coriolis effect until it takes a mostly easterly path around 30° north latitude and at an altitude of about 15 kilometres (9 mi.). The subtropical jet stream may aid or deter the development and steering of tropical storms and disturbances.

The Polar Jet Stream

The northern polar jet stream (also called the polar jet, the midlatitude jet stream, or just the jet stream) has the most influence on weather across much of Canada and the United States and thus receives the most attention on local and national weathercasts. The polar jet stream can usually be found somewhere in the latitude belt from 40 to 60 degrees at an altitude of 7,600 to 10,600 metres (25,000–34,750 ft.). The highest wind velocity is found in the jet stream core, where speeds can be as high as 460 km/h (286 mph) in the winter, but the core typically averages 160 km/h (99.4 mph) in winter and 90 km/h (56 mph) in summer. The segments within the jet stream where winds attain their highest speeds are commonly known as *jet streaks*.

The polar jet stream position marks the location of the strongest temperature contrasts between polar and subtropical latitudes on Earth's surface. Therefore, the strongest polar jet stream velocities usually occur during the winter months. During the summer months, when the equator-pole surface temperature differences are less dramatic, the jet winds blow more slowly and are usually found at higher latitudes. (Though the polar jet streams form along the polar front in the Prevailing Westerlies belt that circles the planet, they do not always circumnavigate the globe as a continuous stream.)

Jet Stream Formation

The polar jet stream forms in the region of the *polar front*, where cold, dry polar air meets the warm, moist air from the subtropical regions. From a climatological viewpoint, the position of the polar front forms a more or less even band around the globe, slipping north and south with the seasonal changes. In the vicinity of the polar front, the air pressure drops more rapidly with increasing altitude in the denser cold air than in the less dense warm air. This temperature effect on air density results in air pressure at any given altitude being higher on the warm (equatorward) side of the polar front than on the cold (poleward) side. When cold air and warm air masses sit side by side, the higher the altitude, the greater the pressure difference between the cold and warm air. Thus, across the polar front, the horizontal pressure differential (or gradient) causes air to flow from the warm side of the front toward the cold side.

Once the air begins to flow, it is deflected by Earth's rotation and

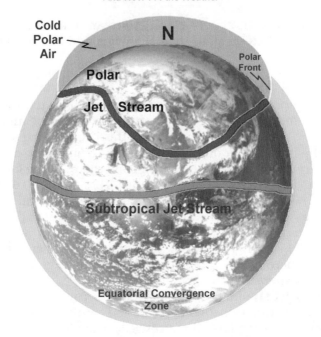

The polar jet stream meanders down over eastern North America, whereas the subtropical jet stream flows zonally across the subtropics.

is prevented from flowing directly from high to low pressure and is deflected to the right in the northern hemisphere. The resulting flow produces a westerly wind current generally flowing toward the east, parallel to and above the polar front. Very strong temperature and pressure gradients in the polar frontal zone can intensify these wind speeds to more than 90 km/h (56 mph), forming a jet stream.

Unlike the climatological position of the polar front, its day-to-day position and shape can vary in latitude and develop prominent north-south loops in the frontal boundary. The streambed of the jet stream can vary widely with these frontal position changes and kinks much like a riverbed develops serpentine bends and meanders. When the flow is mostly west to east, the jet stream is said to be in *zonal flow*. But the jet can take wild north-south swings, a condition termed *meridional flow*. As a result, the position and meanders of the polar front and jet stream are of major importance in defining the weather patterns and conditions over the globe.

Jet Stream around the Polar Front

The polar front and its associated jet stream have a major influence on regional weather conditions. Many surface storms form along the polar front near the jet stream's maximum winds. Due to the association of the jet stream with the polar front, media weather maps often include the jet stream's position as a rough indicator of continental divisions between warm and cold air masses. Here is why.

The average polar-front position slips north and south with the seasons in response to the annual hemispheric heating cycle. The polar frontal belt retreats poleward in the summer months and then expands toward the subtropics during the winter. From a meteorological viewpoint, however, the polar front is not an evenly encircling latitude belt but an undulating zone around the globe that is ever-changing as masses of warm and cold air push away from their regions of origin. Of particular interest to meteorologists and weather forecasters is the pattern of *long waves* that forms on the polar front. These long-waves, very visible on polar projection maps, undulate around the hemisphere with three to six wave cycles.

At times, that global belt fits tightly, having three or four small-amplitude undulations (little north-south latitude variation) around the hemisphere. At other times, it has as many as six large-amplitude loops (great north-south variation). How those frontal wave loops sit over the hemisphere, or portion thereof, determines what temperature regimes are experienced on the surface below. When a long section of the polar front, as seen on continental weather maps, is smooth like the surface of a calm sea, the upper-level winds, including associated jet streams, run generally parallel to the latitude lines (zonal flow). Under zonal flow, north-south undulations of the frontal boundary are small, and the surface temperatures across the continent, as seen in the *isotherm* (lines of equal temperature) pattern on the weather map, layer in zonal (east-west) bands with warm air to the south and cold air to the north.

But, like occupants of a small boat riding that calm surface, we may be lulled during persistent zonal flow by a stretch of nearly indistinguishable day-to-day temperature changes, often remaining near the climatological mean for that time of year. Despite the unfounded desires of those living in the midlatitudes to believe there is a truly "normal" state of weather from which departures are abnormal, such stable zonal patterns are but a temporary state. We live not in the tropics, where day-to-day weather patterns are very similar, but in a region subject to very changeable weather regimes. Thus, when a zonal flow pattern cuts off the exchange of heat between the polar region and the tropics, great thermal contrasts develop across the polar front. This becomes the zonal pattern's downfall. Eventually, some kink, some small perturbation, develops in the zonal pattern, and a *short wave*, with initially small north-south extent, arises upon the polar front. If the short wave *amplifies* (grows in size), it can distort the flow into a new pattern that crosses the latitude lines, a pattern termed meridional. In meridional flow, cold air rushes southward while warm air streams northward.

As the polar-front meridional pattern rises into long waves of deep north-south extent, it severely rocks our "temperature boat." When the wave and associated jet stream plunge southward, we find ourselves in a deep *upper-level trough* with temperatures quite cold below it. When we rise high on the *upper-level ridge*, or crest of the passing long wave, the surface experiences warm, even hot temperatures. Often, the intense temperature changes brought by meridional flow are expressed by frontal patterns associated with a surface low-pressure system: warm air moving north behind the warm front; cold air descending south behind the cold front. Not only are cyclonic storms spawned in the vicinity of the peak jet stream winds (jet streaks), but they are often pushed rapidly eastward by the jet stream above.

When you look at a weather map with the polar jet stream position drawn on it, you can generally tell what type of weather you will be having. If north of the jet, it should be relatively cold, if south, warmer conditions should prevail. If the jet stream flows overhead, look for stormy, quickly changing weather to dominate.

Fifty-five

Solar Winter: Coming of the Dark Days

We know they are coming when we "fall back" from daylight saving time: the *Dark Days*. Though it was good to see the sun rise earlier in the morning, the now-earlier setting of Ol' Sol hits us like a ton of bricks. We leave school, the office, the factory, the mall at the end of the day, and it is already dark or rapidly becoming so.

Before the advent of clocks and our dependency on calendars, cultures recognized the coming of the dark days of the year through observances and celebrations of the last cross-quarter day before the winter solstice. This cross-quarter day marks the entry into solar winter—the darkest annual quarter—for residents of the heavily populated regions outside the tropics and the subtropics, where days are also shorter, but not by a great amount. At Key West, Florida, for example, the midwinter day length bottoms out at about ten and a half hours. North of 76° latitude, night lasts through the full quarter.

This cross-quarter day has gone by many names. The Saxons called it Winter's Eve; the Celts, Samhain. Many rural European societies observed it as the start of a new year. It has been celebrated as the Day of the Dead, and as All Saint's Day in Christian traditions. But today, the greatest observances are held in North America, trailing only Christmas in personal spending on its celebrations. Yes, I'm talking about Halloween.

The date also signifies the beginning of the domination of winter weather conditions in many parts of non-tropical north. The first frigid polar air masses have left the nest and begun their sojourns out of Arctic lands. And so begins the arrival of winter. The active storm centers in the north Pacific and Atlantic now begin storm production in earnest. The Dark Days of the year have begun. The total hours of potential daylight will continue to slowly decrease for another seven weeks or so before bottoming out at the winter solstice, the shortest day of the year, before again slowly increasing. The Dark Days, however, will not end until February, when solar spring returns.

Fifty-six

Gulf of Alaska Storms

As summer wanes and the sun's influence on high northern hemisphere latitudes diminishes, the prime global breeding grounds for northern winter storms renew their vigor. One can be found in the North Atlantic Ocean between the coasts of Greenland and Iceland, the appropriately named *Icelandic Low*. The second lies in the North Pacific off the coast of Alaska between the Aleutian Islands and the southern Alaska coast, which goes by the names *Aleutian Low* and *Gulf of Alaska Low*. These low-pressure zones are semipermanent features of the global weather map, arising from the general circulation patterns. They diminish in extent during the warm season and intensify during the cold months.

The Aleutian Low and the North Pacific High found to its south have a cyclic relationship during the course of the year that directly impacts the weather along the west coasts of Canada and the United States and into the northern portions of the Western Cordillera, the "spinal column" of mountain ranges across the western Americas from the Arctic to Cape Horn. Since this is also a prime source region for Pacific maritime air masses, the weather patterns across the continent are affected. Within the subpolar zone, warm air moving north from the tropics and cold air moving south from the Arctic clash, a conflict that is strongest during the cold half of the year and weakest during the warm half. After an active winter, the Aleutian Low diminishes slightly in March and by April is sending fewer storms outward. The feature virtually disappears in July, appearing as a weak trough of low pressure rather than a tight low-pressure cell. As solar energy falling on the Arctic regions diminishes in September, the Aleutian Low again strengthens, ready to produce a litter of winter storms.

The clash between cold and mild air generates spinning low-pressure cells that grow to energetic youth over the Gulf of Alaska.

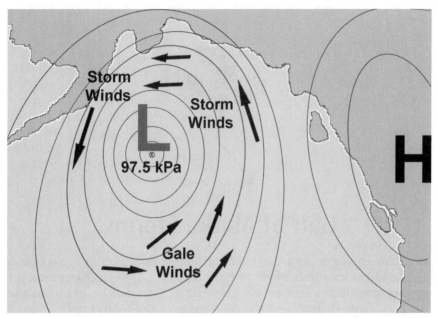

A typical Gulf of Alaska storm pattern off the Alaska–British Columbia coast.

As a result, meteorologists call the region a "center of action" for North Pacific maritime and coastal storms. The Aleutian Low region is characterized by barometric pressure readings of 101.3 kilopascals (1,013 mb) or less, high humidity, and generally complete cloud cover. When a newborn storm spins up, the region experiences strong winds and heavy precipitation, most falling as rain despite the high latitude. By November, the Aleutian Low average surface pressure has dropped to 100.2 kilopascals (1,002 mb), a level it maintains for much of the winter.

The Gulf of Alaska lies in the arched bite of the state, south of Anchorage between the Kenai Peninsula and Kodiak Island and the eastern Alaska Panhandle. The gulf's northern coast is fortressed by the Wrangell/St. Elias Mountains—one of the world's highest mountain ranges, still rising at the geological hyperspeed of 4 centimetres (1.6 in.) per year—which effectively block most storms from entering eastern Alaska and the Yukon. The Gulf of Alaska is the most common entry point for the northeastern Pacific storm track that sends stormy weather to the North American coast.

Storms move out of the Gulf of Alaska into the Pacific Northwest region of the continent with great regularity, as many as three or four per week during the height of winter storm season. Quite often, these storms are packed so closely that the Pacific Northwest does not experience a long break of clear weather, so common over the rest of the continent, between them. I have compared this train of tempests to the passage of a nearly neverending freight train of boxcars. As each storm rolls by, the clearing break is often no more than a tease, like the daylight seen between that rolling chain of boxcars. Only rarely in some areas of the coast do "flatcars" appear—bringing longer clearing—and the "caboose" may not arrive until spring.

Wintertime Gulf of Alaska storms typically sport winds in excess of 80 km/h (50 mph), with corresponding high waves. Some storms rival the strength of great hurricanes and typhoons and would be considered as such if their origins were tropical rather than polar. (In fact, the U.S. Hurricane Hunter Air Squadron often deploys aircraft into these winter storms to study their structure.) Winds as high as 160 km/h (100 mph) have been recorded, with wave heights pushing 20 metres (66 ft.) or more. Over the years, these storms have taken their toll on fishing and merchant vessels caught in open waters.

When these storms hit the North American coast, they encounter a series of high mountain ranges paralleling the coast from central Alaska south into northern California. The mountains force the storm air to great altitudes, causing them to dump much of their prodigious moisture content on the coastal mountains as rain and snow. The result is the incredibly dense temperate rainforests that cover the coast from Alaska's Panhandle through British Columbia and south into California. In these regions, rain and snow are measured in metres annually. During the winter, nearly all days report rain or snow somewhere in the region. The wetness that sits in the region can be exemplified by Prince Rupert, British Columbia, with 240 wet days annually; Juneau, Alaska: 222 wet days; Clearwater, Washington: 192 wet days; and Astoria, Oregon: 190 wet days. The world's record for snowfall in one *snow year* (which runs from 1 July to 30 June) was established on Mt. Baker in Washington in 1998–99 at 2,986 centimetres (nearly 30 metres or close to 98 feet). North America's wettest year, 1976, drenched MacLeod Harbor on Montague Island, about 112 kilometres (70 mi.) east of Seward, Alaska, with 8,440 millimetres of rain (that's 8.44 metres or about 28 feet).

Fifty-seven

Jack Frost's Fingers

"He comes, he comes, the Frost Spirit comes!"
(John Greenleaf Whittier, *The Frost Spirit*)

When I think of November, the first visions that pop into my head are those of cloudy and dark days. In each region of North America where I have lived, the low sun and short days of November enhanced the gloom of skies frequently flushed with stratus clouds from dusk to dawn. But often November mornings dawn quite the opposite, sporting a quiet, brilliant beauty. These are the mornings touched by Jack Frost's brush.

The word *frost* can take any one of several related meanings in the English language. To farmers, gardeners, and botanists, it means those subfreezing temperature conditions that halt or destroy plant functions. A frosty day is generally considered to be any day when temperatures are subfreezing, regardless of other weather elements. *Degrees of frost* refers to the number of temperature units below freezing (0°C [32°F]). Finally, the aspect of frost I will discuss here: "frost" refers to ice crystals appearing on non-liquid surfaces without the aid of precipitation, most often forming during the night.

Frost, the cold cousin of dew, usually appears during clear, cold nights with light winds when the air temperature near a surface falls below the frost point, the temperature to which air must be cooled—at constant pressure and humidity—to achieve saturation with respect to ice at or below 0°C (32°F). Frost can even form when the officially reported air temperature is above freezing. For frost to form, it must have a surface with a temperature below the frost point. Thus, frost may form on rocky, glass, or metal surfaces that lose heat more rapidly than the surrounding air through radiative cooling. This is why car windshields can frost over yet no frost forms on the pavement or vegetation surrounding the vehicle. Frost formation may also be sporadic around an area, particularly in hilly or mountainous terrain. Since cold air is denser than warm air, it flows like water downhill to pool in low areas or hollows. Low-lying areas more susceptible to frequent frost formation are called *frost pockets* or *frost hollows.*

Frost can form in one of two ways, both similar in nature to the formation of dew but ultimately leaving solid water—ice—on the surface rather than liquid water. At temperatures from about 0 to –18°C (32 to 0°F), frost will most likely initially form when water vapor condenses as a supercooled liquid on a surface and then quickly freezes. Once the first ice crystals have formed, additional frost accumulation may proceed by the deposition of water vapor directly to the solid (ice) state. On certain surfaces, frost can begin to be deposited directly without condensation and freezing to form that initial ice crystal. In such cases, some characteristic of the surface or impurities on it can seed the process by mimicking the structure of an ice crystal to provide a growth site. This can be a surface flaw such as a scratch on a piece of glass or a chemical impurity or particle lying on its surface. For example, crystals of silver iodide so closely mimic the structure of ice crystals that they are used for condensation nuclei to *seed* clouds in rain- and snowmaking endeavors. At colder temperatures, ice will form directly on the surface through the accretion of water vapor directly from the solid state. This process is called *deposition*. (Some incorrectly call it *sublimation*, but technically that is the opposite process: the conversion of ice directly into water vapor without an intervening liquid phase.)

There are two forms of frost: *rime* and *hoar*. Rime frost occurs when the rate of frost formation is rapid, usually under conditions of high water content in the air (vapor and/or liquid) and at least moderate wind speeds. Rime formation is common during fogs where supercooled or near-freezing water droplets come in contact with subfreezing surfaces. It may also form when clouds are blown over subfreezing surfaces, such as occurs when moisture-laden clouds are forced over coastal mountains during the cold season. In these cases, rime accumulations can be several metres thick, growing

Winter rime encases the observatory atop Mount Washington, New Hampshire. (Bryan Yeaton)

in the direction of the prevailing wind at the time of formation. Rime frost, on close inspection, has a grainy appearance, like sugar or salt. Rime is opaque, less transparent than glaze ice formed during freezing rain episodes. It is denser and harder than hoar frost.

Hoar frost forms from the slow deposition of water vapor more or less directly to ice on a surface: Jack's frost. By accumulating slowly, hoar frost can form interlocking crystals that grow out from the surface. Hoar frost generally has a feather, fern, or flower pattern growing from the initial seed. It forms only when winds are light, which is often the situation during clear, cold nights; thus it is the most common type of frost in non-mountainous areas. When we speak of frost, we usually mean hoar frost rather than rime frost.

Hoar frost shows a definite, delicate crystalline structure of elements growing on elements in steps or layers. The crystal's white color is caused by small air bubbles trapped in the ice, thus reducing its transparency. The smooth faces of the hoar crystals allow it to glitter in the sunlight, particularly at the low sun angles of early morning. This appearance is in sharp contrast to rime frost, which adds a dull, matte finish to the surface on which it adheres. Hoar frost crystals are usually small, but under the proper conditions, if left undisturbed, frost ferns and frost flowers may grow quite large. In the quiet conditions found in an ice cave or ice-filled sinkhole, for example, they may rival true ferns and flowers in size.

When frosts spread across the countryside, the landscape wears a veneer of dazzling beauty as the sunlight is scattered, reflected, and refracted by the many crystal surfaces. Such frosty November mornings reveal the nightshift labors of Jack Frost, bringing much-needed beauty to a month generally characterized in northern latitudes by dull, dark days. And while frosts provide great beauty on the large scale, they are best appreciated close up. The delicate crystal patterns of hoar frost can make even the most common surface a treasured vision.

I heartily recommend, the next time you wake up to a frosty morning, that you take a few minutes to study a variety of frosted objects through a magnifying glass—it will open a new world of intricate natural beauty. This is one time when your weather eyes should be looking *down*.

Frost produces a delicate crystal pattern on a glass window. (KEITH C. HEIDORN)

Fifty-eight

The Winds of November

As solar winter descends upon North America in early November, much of the continent also experiences increased cloudiness as large storm systems move across the continent. November rightly deserves the title of Gloomiest Month of the Year. Gloomy skies are not the only hallmarks of November—windiness and storminess also characterize the month. The Anglo-Saxons called the eleventh month of the year the *winde-monath*. In many areas of North America, particularly along the Canada–U.S. border, the name would be appropriate as well. In November, the first of the strong winter storms follow a storm track south from the polar regions into North America's most heavily populated areas. These storms begin their annual trek just about the time the Atlantic tropical storm season sounds its all clear.

As autumn blends into winter, North America is besieged with storminess. In the northwest, the Aleutian Low slips into its winter position in the northwestern Gulf of Alaska. This mother of storms sends wave upon stormy wave against the Pacific coasts of Alaska, British Columbia, Washington, and Oregon. High winds and waves accompany these storms on the seas. When the storms collide with the coastal mountains, heavy and steady rains fall, with snow at higher elevations. Across the continent off the northern Atlantic coast, powerful Nor'easters form. These storms can lash the coastline and offshore waters for days with a fierceness equal to hurricanes. To make their signature unique, these storms add ice, snow, and frigid temperatures to the deadly mix of wind and rain, wave and storm surge.

Great Lakes Storms

Over the largest bodies of fresh water in the world, the Great Lakes, two storm tracks converge in November. One brings storms south

Christmas Snowstorm
GOES-8 Visible (4km)
December 25, 2002 @ 1845 UTC

A Christmas snowstorm in 2002, extending from the Great Lakes to the Atlantic Coast, is captured by a camera on the GOES-8 satellite. (NATIONAL OCEANIC AND ATMOSPHERIC ADMINISTRATION/U.S. DEPARTMENT OF COMMERCE)

from Alberta, the precursors to the *Alberta Clipper* blizzards. The second track shoots storms from the lee of the Rockies near Colorado, known as *Colorado Lows*, toward the Great Lakes basin. When these storms cross the region, the waters of the Great Lakes often put an even more deadly spin on the cyclonic systems.

The winds do blow. Storm winds frequently exceed hurricane force on the open waters. In fact, mariners have called such powerful autumn storms *Great Lakes hurricanes*. And the lake waves do pound. Strain gauges mounted on the SS *Edward L. Ryerson*, a 240-metre (787-ft.) ore carrier, measured a stress of 23,000 pounds per square inch (159 MPa) on its hull during a November 1966 Lake Huron gale. Though the *Ryerson* survived, the 256-metre (839-ft.) freighter SS *Daniel J. Morrell* broke in two in that gale and sank in 61 metres (200 ft.) of water. Wind and wave are not the only weapons of the "November Storm King." Great Lakes mariners must also face driving rain or snow, ice, and freezing temperatures. Although storms can sweep across the Great Lakes region during any month, November is the month most honored in memory and song. What is the reason for this infamy?

The Great Lakes play a major role in determining the climate and weather of their region. The reason for their large influence lies in their waters. Water gains or loses heat much more slowly than air or land surfaces. Thus, the large volumes of water in the Great Lakes cool so slowly that the water temperatures of all the lakes are out of step with the seasons by a few months. As autumn progresses, lake waters retain much of their summer warmth. When the first cold, northern air masses move out of the Arctic and across the lakes, they are warmed by the waters below. This added heat tempers the arctic outbreak, postponing the first frosts along the southern and eastern lakeshores by several weeks. The fruit belts of Ontario, Michigan, and upper New York State are made possible by this moderating influence on autumnal cold air outbreaks.

By mid-autumn, the contrast between the cold, dry air moving down from the Canadian North and the warm, moist air flowing up from the Gulf of Mexico can cause great storm systems to form along the polar front. Many such storms track toward the Great Lakes basin in November. As these storms move across the lakes, they receive heat energy from the warm waters below, which provides additional fuel to run the storm engine. When a developed or developing storm system moves across the relatively warm waters, it can intensify with explosive speed. The resulting storms may produce hurricane-force winds reaching 160 km/h (100 mph), large waves that at times exceed 15 metres (49 ft.) in height, and heavy precipitation over the water and along the shoreline. A few storms stall over the Great Lakes basin, voraciously feeding and growing on the warm-water energy below them. While spinning in place, such storms may ravage the lakes and surrounding shoreline for days.

Many great storms have been born over the lakes during November, and their toll on Great Lakes shipping has long been the subject of story and song. At least twenty-five killer storms have greeted November sailors on the five lakes since 1847, bringing death to more sailors on the Great Lakes than any other agent.

Ships caught on the Great Lakes during such fierce storms can be tossed like toys in the fury of wind and wave. As early as 1835, a November storm "swept the lakes clear of sail." In 1847, a major storm claimed 77 ships on the Great Lakes. Ten years later, 65 vessels went down as a storm crossed the lakes. A 1905 gale on Lake Superior wrecked 111 ships and sent 14 steel carriers ashore. In 1958 and 1975, powerful storms also caused shipwrecks and damage over the Great Lakes.

The king of Great Lakes storms, however, struck Lakes Huron and Erie in 1913, when wind and wave sent 12 vessels to the bottom at a cost of more than 230 lives (some estimate as many as 300) and pushed 20 additional ships aground. The winds on Lake Huron on 9 November 1913 blew from the north at speeds steadily exceeding 110 km/h (68 mph), with gusts in excess of 139 km/h (86 mph). As the storm, known historically as the "Ultimate Storm" and the "Big Blow," moved across Lake Erie, the barometer dipped to 96.9 kilopascals (969 mb) in Erie, Pennsylvania. Buffalo, New York, measured winds of 100 km/h (62 mph), and 56 centimetres (22 in.) of snow fell on Cleveland, Ohio, after a day of freezing rain. The winds blowing across Lake Erie from the west caused the lake level to drop by up to 1.8 metres (5.9 ft.) in the western basin, leaving several boats sitting on the muddy bottom of Maumee Bay, Ohio. As the storm moved across Lake Erie, it blasted Ohio residents with wind and snow and cold temperatures.

The beat of November storm wind and wave has been heard by a good many Great Lake mariners. Among the many ships lost on the Great Lakes waters when the storm winds of November blew deadly were the schooner *Black Hawk* (1847), the schooner *Persia* (1869), the ore carriers *Charles S. Price* and *James B. Carruthers* (1913), the steel freighter *Novadoc* (1940), the limestone carrier *Carl D. Bradley* (1958), and the ore carrier *Edmund Fitzgerald* (1975), which was later immortalized in song by Canadian folk singer Gordon Lightfoot.

Mariners still face the Great Lakes and their angry, stormy moods, but modern communications, advanced weather forecasts, and radar now provide some degree of warning when November's angry gales come slashing.

Fifty-nine

Clippers and Nor'easters

The great tropical storms roaming the world's oceans and seas have earned the right to be given individual names: Andrew and Hazel, Keli and Jelawat, to name but a few. Extratropical cyclonic storms, those forming from the clash of polar and tropical air masses, have not yet been honored by regular naming. (Though many believe the naming of hurricanes began with George Stewart's classic novel *Storm*, in which the junior meteorologist names cyclonic extratropical storms that formed over the Pacific Ocean.) A few, by virtue of their devastation or magnitude of snow- or rainfall, earn names such as the President's Day Storm, the Blizzard of '88, and the White Hurricane. However, some extratropical storms form so frequently in specific areas or have such particular qualities that we often recognize the genre with a unique name.

Two common North American families of cyclonic entities are the Alberta Clipper, which races across the continent out of the Alberta plains, and the Nor'easter, which races up the Atlantic coast.

The Alberta Clipper

Many major extratropical storms are born in the lee of the Rocky Mountains in two prime mid-continent spawning grounds: one around Colorado, the other in Alberta. Those systems born near Colorado are simply called Colorado Lows. Storms born in Alberta (perhaps *reborn* is a better concept because many arise from remnants of Gulf of Alaska lows weakened in crossing the coastal mountains and the Rockies) possess unique characteristics that have led to the handle Alberta Clippers because they move swiftly out of the Canadian prairies, like the clipper sailing ships of old, bringing frigid weather to the east.

Alberta Clippers form on Alberta's high plains east of the Canadian

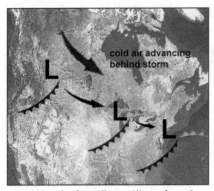

cold air advancing behind storm

Typical path of an Alberta Clipper from the foot of the Rockies to the Atlantic Coast.

Rockies. Once organized, they "sail" southeast under the push of a northwesterly jet stream into Montana and the Dakotas, then turn across the Great Lakes basin, headed toward the Atlantic coast. This track, however, leaves them far from the moisture sources of the Pacific Ocean, Atlantic Ocean, and the Gulf of Mexico. Due to their high forward speed and relative dryness, average Clippers rarely deposit huge snowfalls, dropping instead only a few centimetres of light, powdery snow in passing. That doesn't mean, however, that Alberta Clippers are insignificant storms. At full maturity, a Clipper's cold winds may reach 65 km/h (40 mph), with gusts to 100 (62 mph). Although the amount of snowfall may be small, these strong, frigid winds produce severe blowing and drifting of any snow on the ground, causing true blizzard conditions along the storm's track.

When an Alberta Clipper passes, it often ushers in bitter outbreaks of polar air, which continue the storm's wrath for days after the low itself has moved eastward. Combine strong northerly winds and bitterly cold temperatures in the storm's wake, and you get severe ground blizzards, days of *whiteouts* (severe blowing and drifting snow reducing near-surface visibility to zero), and dangerous windchill conditions. When frigid, gusty northwest winds follow in an Alberta Clipper's wake across the Great Lakes, they may bring major lake-effect snowfalls along the lee shores, often causing heavy snowfall over the region. Occasionally, an Alberta Clipper slips farther south to cross the Appalachians and southeastern American coast. There, when conditions are ripe, the waning storm becomes re-energized by the warm Atlantic coastal waters to be reborn a Nor'easter.

The Nor'easter

In all cyclonic flow patterns, winds spiral counterclockwise around the low-pressure center in a push-and-pull struggle between pressure gradient, Coriolis, and frictional forces. In a well-developed cyclonic storm, the flow regimes develop sectors with strong northeasterly, southwesterly, and northwesterly winds. Therefore, any major cyclonic storm has the potential to be called a *southwester*, a *northwester*, or a *northeaster*, depending on the origin of its strongest winds over an area. (Actually, such storms may be all three at once in regions hundreds of kilometres apart.) However, only along the North American Atlantic coast, from Cape Hatteras north to the Canadian Maritimes and Newfoundland, can a storm truly be called a Nor'easter. By accepted definition, Nor'easter winds blow at gale force or stronger, coming initially from the northeast. These storms often bring heavy precipitation, falling as rain, snow, or, at times, freezing rain. Along the coastline, heavy surf generated by the off-shore storm winds may cause extensive damage.

Nor'easters are the most common, widespread severe weather events to worry Maritimers and New Englanders. The term "Nor'easter" arose in the colonial days before scientists established the concept of the wind circulation pattern around a low-pressure center. During that time, storms were believed to travel from the direction of the wind. Thus, if a storm had strong southwesterly winds, it had arrived from the southwest of the observer. A Nor'easter, therefore, originated to the northeast of those feeling its fury.

Nor'easters have become legendary along the Atlantic coast for their fury and impacts. The Blizzard of '88 and the Perfect Storm of 1991 are two examples, as was the massive Nova Scotia snowstorm of February 2004, dubbed Winter Juan (the cousin of Hurricane Juan of that autumn). Most Nor'easters arise from cold-season storm systems that first move through the American Ohio Valley or Gulf states to a position off the Atlantic coast (often beginning as Alberta Clippers, Colorado Lows, or Texas Lows). When they encounter the warmer Atlantic waters, many undergo a regeneration of intensity, particularly if Gulf Stream waters are involved. When conditions are ideal, the storms become *bomb cyclones*, a term relating to a sudden and dramatic deepening of the central pressure—at a rate equal to or exceeding 1 millibar per hour over a twenty-four-hour period. The suddenness of a bomb cyclone's fury is matched only by the resulting winds and precipitation, which can rival hurricane intensities.

Many Nor'easters hug the coast as they move north. Thus, their northeast storm winds rush off the ocean with unbridled fury. Waves and storm surge generated by these winds may cause devastation along the coastline. The wind-flow pattern also allows the storm to pick up relatively warm and very moist air from the ocean, which it drops when meeting colder air on the system's northern side. If the air is cold enough, the precipitation falls as snow—merging blizzard with Nor'easter.

Lake-Effect Snowfalls

During the late autumn and winter, when cold arctic air sweeps across the Great Lakes, snow squalls may form along the lakes' lee shores. These squalls can bring heavy snowfalls with reduced visibility to a relatively small area. Often, while squalls hit one area, blue skies prevail several kilometres away. At other times, their impact can cover kilometres of coastal regions and reach far inland.

Most lake-effect snowfall occurs, not during the passage of the low-pressure cell of a winter storm, but in the strong cold airflow behind the storm's cold front. In areas south of the Great Lakes, cold front passage will often be followed by a twenty-four- to thirty-six-hour period of blustery northerly or westerly winds, falling temperatures, and persistent flurries of fluffy snow. Heavy snow squalls accompanied by blowing snow and reduced visibility mix with brief periods of partly cloudy skies and some blowing snow.

When lake-effect snow squalls are well developed, there may be

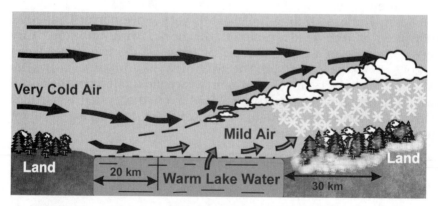

Idealized formation pattern of lake-effect snow developing over the Great Lakes.

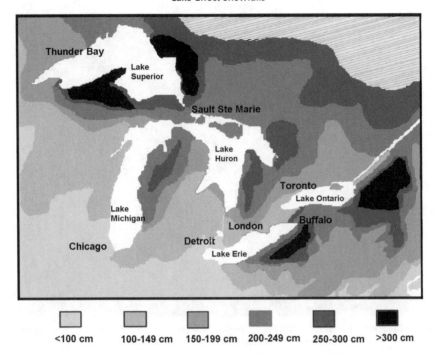

Annual average snowfall (in centimetres) around the Great Lakes basin showing regions of greatest lake-effect snowfalls.

less than twelve hours between the last of them and the more widespread snowfall of the next approaching cyclonic storm system. Thus, lake-effect snows often add to the miseries of a winter storm's passage. When cyclonic snowstorms pass through areas such as the prairies/Great Plains, the advancing cold arctic air will quickly clear skies. Although in these post-storm situations, temperatures are cold and, if winds are strong, fallen snow can blow or drift, the snowfall is usually over; not so around the Great Lakes, where it is said, "During the winter, the weather clears up stormy." The cold air picks up substantial moisture as it moves over the lakes and deposits it as snow inland from the downwind shore. Often, accumulations exceed those deposited during the storm itself. In the more severe lake-effect snow squalls, snowfall accumulations of more than 75 centimetres (30 in.) per day are not uncommon, and fall rates as high as 28 centimetres (11 in.) per hour have been reported. Such severe snowfalls have been termed *snowbursts*.

Lake-effect snows are not restricted to Great Lakes shorelines, but are most common and heaviest there. Any large lake may produce lake-effect snow downwind if it remains essentially ice-free and its waters warm relative to the overriding cold air. And, although the

Snow streamers (parallel cloud lines lying NW-SE) form over the Great Lakes on 16 January 2004. (NATIONAL OCEANIC AND ATMOSPHERIC ADMINISTRATION/U.S. DEPARTMENT OF COMMERCE)

east and south shores of the Great Lakes are most likely to have lake-effect snowfalls, north and west shorelines can be hit if the air traversing the lake is cold enough and the distance the air passes over open water (the fetch) is large enough.

What Causes Lake-Effect Snow Squalls?

Lake-generated snow squalls form when cold air, passing for a long distance over the relatively warm waters of a large lake, picks up moisture and heat and is then forced, upon reaching the downwind shore, to drop the moisture in the form of snow. Lake-effect snows are common over the Great Lakes region because these large bodies of water can hold their summer heat well into the winter, rarely freeze over, and provide the long fetch that allows air to gain the heat and moisture required to fuel the snow squalls. Lake-effect snows are most pronounced and effective wherever terrain features such as small hills or mountains are oriented along the lee shores, as in northern New York State.

Each year, as the warm weather of early autumn fades and winter's first cold blasts rage, the waters of the Great Lakes become increasingly warmer relative to the cold air masses formed in the north. When cold and relatively dry air traverses the lake, the lower

244

levels are warmed and moistened. This air now becomes lighter than the air above it (a condition known to meteorologists as *convective instability*) and starts to rise rapidly. Depending on the degree of the air mass's instability (that is, how much warmer the lake water is than the air), bands of stratus, stratocumulus, or heavy cumulus clouds will form over the water, advancing with the wind toward the downwind shore.

The upward motion of air continues as the air mass moves toward the lee shore, where the land surface slows down the onshore flow, creating a piling up, or convergence, of air. The air, having no place else to go, rises rapidly, triggering the formation of snow showers or squalls. Sloping terrain along the shoreline also enhances vertical motions. If snow showers have not begun to fall over the lake, they soon fall as the air rises over the shoreline. The intensity of the lake-effect snowfall depends upon several factors: the temperature contrast between the lake surface and the air passing over it, the over-water fetch, and the regional weather situation. The distance snow squalls travel inland increases under higher wind speeds, while their direction is controlled by the winds flowing above the surface. A snow squall's maximum penetration inland will generally be greatest during late autumn and early winter and shortest during late winter.

Most often, lake-effect snowfalls take the form of light to moderate flurries spread over a broad but still limited area. However, an individual squall of heavy snow may remain over one small area for several hours and then, with a shift in the wind direction, move to drop its snow on another area. Satellite and radar observations show that lake-effect snow clouds most often occur in bands resembling streamers. These bands usually form over the lake and are swept inland by the winds. In the Great Lakes, small multiple bands appear to develop when the wind blows across a lake's width. However, if the wind blows across the length of a lake, a single large cloud band as wide as 80 kilometres (50 mi.) and 40 to 160 kilometres (25–100 mi.) long may form. Such intense cloud bands cause highly localized blizzards with swirling, blowing snow, reducing visibility to zero.

Lake-effect snow cloud bands are remarkably persistent. They have been observed to cause continuous snowfall for as long as forty-eight hours over a sharply defined region. One single intense local storm cell can yield as much as 120 centimetres (47 in.) of light-density snow in twenty-four hours or less. As a result, winter weather in the lee of the Great Lakes shows a complex variability of snowfalls, with areas of deep snowfall often adjacent to areas with relatively little snow.

Sixty-one

A Hard Rain

When you mention winter precipitation to most North Americans, the first image that generally comes to mind is one of snow. But often along the winter storm fronts, where the forces of warm and cold air clash, precipitation may fall in a variety of forms other than snow or rain, even if only briefly. Rain and snow, you see, are not the only members of the winter precipitation congregation, just the most common. The other winter members include *freezing rain* and *drizzle, hard rime, ice pellets, ice crystals, ice needles, snow grains,* and *graupel.* They are also known by a variety of local names in different regions: sleet, hail, soft hail, snow pellets, and diamond dust.

Hard rains may fall either as showers or continuous precipitation, but some forms fall in a preferred manner. Like snow and rain, ice pellets may fall continuously or in showers. Freezing rain/drizzle, snow grains, and ice crystals usually fall as steady precipitation, whereas showers are the dominant method of fall for graupel—snow pellets and soft hail. Soft rime generally deposits on surfaces rather than falls. Almost all forms of hard rain begin as ice or snow crystals in clouds. If, as they descend, the crystals traverse a layer of air with above-freezing temperatures, they can melt or partially melt, depending on the depth of the layer and its departure from freezing. If they do not again enter a subfreezing air layer, they fall as rain, or a mixture of rain and snow. But if they pass through another layer of subfreezing air before hitting Earth's surface, they may refreeze, partially refreeze, or perhaps supercool. And if the falling particles are caught in a series of updrafts and downdrafts and make multiple passes through the warm and cold layers, they may collect several layers of ice like a hailstone.

The two exceptions to the above are ice crystals and ice needles, which are special forms of snow. Ice crystals may form under clear skies and light winds when the temperature drops and fall as soft showers. As they flutter slowly downward, they may glitter in sun- or moonlight, earning the name diamond dust. Ice crystals form at low temperatures (less than –18°C [–0°F]) as hexagonal plates and/or columns. The latter resemble stubby pencils rather than delicately branched snowflake shapes. Ice needles are special varieties of snow or ice crystals. These narrow, pointed ice crystals, 1 to 3 millimetres (0.04–0.12 in.) long, fall when the air temperature at formation is in the narrow range of –4 to –6°C (25 to 21°F). Ice-needle snow may pro- duce high avalanche danger when falling in mountain areas.

Ice pellets or ice grains are clear or opaque hard ice particles, spherical or irregular in shape, with diameters of 1 to 5 millimetres (0.04–0.2 in.). They are formed by the refreezing of rain or drizzle drops or partially melted snowflakes falling through a cold air layer near the surface. Often called *sleet* in the United States, they bounce when hitting a hard surface. Graupel, *snow pellets*, and *soft hail* are soft, opaque ice particles, irregular in shape and often studded with oblong or branched protrusions. Graupel (which, as noted earlier, is also an ingredient in hailstone formation) splatters when it hits a surface. With diameters of 1 to 7 millimetres (0.04–0.28 in.), graupel forms in strong updrafts of cold air in blizzards, lake-effect snows, and thundersnows, and is often highly electrified. Graupel, whose name is derived from the German word for barley, is created when supercooled water droplets coat a snowflake with ice or when a supercooled drop develops an outer coating of ice without freezing through. Since graupel forms in convective updrafts and may be pushed quickly downward in descending currents, it may fall when surface temperatures are significantly above freezing.

Sleet is a term for cold-weather precipitation that has a few dif- ferent meanings. In the United States, it refers to ice pellets, but is sometimes erroneously used to mean freezing rain. In the United Kingdom, sleet refers to mixed rain and snow or melting snow.

Snow grains are white, opaque particles, either flat or elongated, and less than 1 millimetre (0.04 in.) in length, the solid equivalent of drizzle. Smaller and flatter than snow pellets, they generally fall from stratus clouds rather than convective clouds. The final forms of hard rain are *rime* and *glaze*. There are two forms of rime: hard and soft. Hard rime and glaze are more technical subcategories of freezing rain or drizzle (see "Ice Storms: Beauty amid Destruction," page 252). Both are formed by supercooled drops that freeze on contact with a cold surface. Temperatures and rates of fall determine whether freezing rain or drizzle is made up of glaze ice or rime ice. Meteorologists classify transparent and homogeneous ice forming on vertical and horizontal surfaces as glaze. The amorphous, dense structure of glaze helps it to cling tenaciously to any surface on which it forms, causing extensive damage across the region where it falls. In contrast, if the ice is milky and crystalline, it is termed *hard rime*. Hard rime ice is less dense than glaze ice and clings less tenaciously, therefore generally inflicting only minor damage. Soft rime is deposited on cold surfaces by supercooled cloud droplets, rather than rain falling onto it. The buildup of soft rime ice usually leaves an opaque coating, which may point like a flag in the direction of the wind that deposited it. Soft rime is also often called hoar frost (see page 234) and is common at high elevations, where cold clouds move across the landscape.

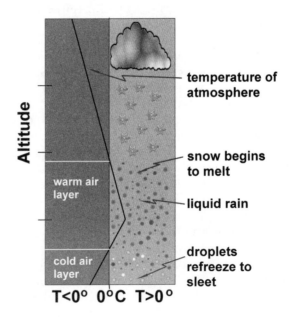

Vertical atmospheric conditions required for sleet formation.

The Chinook

They flow off the mountain ridges, rushing winds that are relatively very warm and very dry. Along the eastern slope of the Rocky Mountains, they have been called *Snow Eaters*, but today are more commonly known in the western prairies and plains regions by their Native American name: *chinooks*. Similar hot and dry winds descending mountain slopes are found around the world. Since they were first studied in the Alps, they are generically known by their local Alpine name: the *foehn* (or *föhn*) winds. Foehn winds flow off mountain ridges, usually in the lee of the prevailing wind direction. In North America, they have had several other local names, including the *Santa Ana winds* of California.

The foehn-wind members are warm and dry winds because their air has been warmed by compression as they leave mountain summits and descend leeward slopes. To describe how they typically form, I'll use the Canadian (Alberta) chinook as the example. From a global perspective, the prevailing wind regime, the Prevailing Westerlies, blows across southern Canada from the Pacific Ocean toward the Atlantic shore. We first pick them up approaching the Pacific coastline, where, in crossing the northern Pacific waters during winter, they have extracted heat and moisture from the underlying ocean surface.

When these relatively warm and very moist winds make landfall, they encounter the coastal mountain ranges of British Columbia and are forced to rise over the mountain barrier on their trek eastward. As the flowing air mass rises toward the summit ridges, it cools until it eventually reaches its saturation temperature. At this time, its water vapor burden condenses into liquid water, first forming clouds, then precipitation. Much of that liquid, or frozen, water eventually falls as prodigious amounts of rain or snow on the Coastal Ranges.

Although the air mass cools by expansion as it rises over the mountains, it gains back a great deal of heat when its water vapor converts to liquid water—*the latent heat of condensation*, which amounts to about 2.5 kilojoules or 597 calories per gram of water. (Additional heat—*the latent heat of fusion*, 0.33 kilojoules or 79.7 calories per gram—is released should the liquid water freeze to ice within the air mass.) By the time the air mass has traversed all of British Columbia's ranges, most of its water content has been lost through precipitation; however, a good portion of that released latent heat still remains in the air mass. When the airflow descends from the high ridges of the last mountain barrier, the Rockies, onto Alberta's high plains, it warms through air compression, like the air in a bicycle pump when the plunger is pushed. The air mass warms by about 9.8 C° per 1,000 metres (5.4 F° per 1,000 ft.) of descent. Since many of the ridge lines in the Rockies are 3,000 metres (9,836 ft.) above sea level and the Alberta plain is around 1,000 metres (3,300 ft.), the air will warm by nearly 20 C° (35 F°) in its descent. The air parcel is also very dry since it lost most of its initial moisture content crossing the mountains while gaining very little new moisture during its journey. This warm descending air is the chinook.

The most impressive chinook winds blowing off the Rockies can reach speeds of between 65 and 95 km/h (40–59 mph), with gusts exceeding 160 km/h (100 mph). When blowing at those speeds, a chinook can tip railcars off their tracks and blow semi-trailer units off the road. Impressive as the chinook is as a wind, the temperature changes it brings can be astonishing, often as much as 20 to 25 C° (36 to 45 F°) in an hour. The greatest chinook temperature jump recorded occurred on 22 January 1943, when a chinook shot the temperature in Spearfish, South Dakota, from a chilling –20°C (–4°F) at 7:30 AM to a balmy 8.3°C (47°F) just two minutes later, a rise of more than 28 C° (50 F°)! In Pincher Creek, Alberta, a chinook jacked the temperature 21 C° (38 F°) in the space of four minutes on 6 January 1966.

No wonder the chinook has the reputation its name "Snow Eater" engenders! The deadly-to-snow combination of high temperature and dry air rushing by at high speeds can literally remove a foot of snow in a few hours. It may not just melt the snow but evaporate it as well, often all in one single process called sublimation, without leaving a liquid pool behind, as the very dry air soaks up the liquid like a sponge, replenishing some of the humidity lost in the mountain traverses. When, on 25 February 1986, a chinook descended on Lethbridge, Alberta, with winds gusting to 166 km/h (103 mph), it fully removed a snow pack 107 centimetres (42 in.) deep in eight hours. Lethbridge was left with substantial wind damage and new lakes standing in the surrounding fields and pastures.

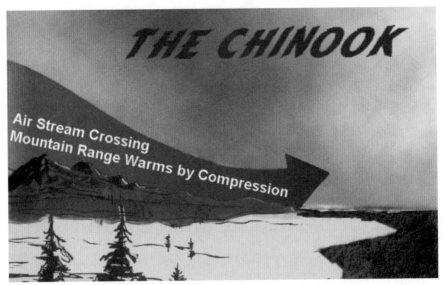

The chinook descends from high mountain elevations.

Chinook winds may last for only a few hours or a few days, but sometimes will persist for several weeks. They can be welcome visitors during the winter, giving a respite to residents of the cold prairies—their version of a January Thaw. For many living under the chinook influence, however, its winds bring debilitating physical effects ranging from restlessness and sleeplessness to anxiety and severe migraine headaches. Recent studies also suggest chinook winds rolling off the mountains can damage aircraft at cruising altitudes by generating turbulence powerful enough to rip an engine from a jet. But in other seasons, the searing dry winds can desiccate vegetation, raise soil into dust storms, and rapidly increase the danger of grass and forest fires. The fire menace of foehn-type winds brings much worry in many areas, particularly in summer and autumn. In the Canadian prairies, concern heightens over grassfire potential during dry periods. Similar worries arise each year between October and February for residents of Southern California. During this period, when the Santa Ana winds blow, the dry chaparral country becomes a tinderbox needing only a spark to touch off devastating wildfires.

Sixty-three

Ice Storms: Beauty amid Destruction

If the blizzard is the king of winter storms, the queen must surely be the *ice storm*. Often within the realm of a single large winter storm, the Blizzard King will sit on the cold seat of power, blustering over his subjects to the west. To the Blizzard King's left, the Ice Storm Queen sits, dispensing destruction encased in great beauty, a dazzling crystal sheath falling over the lands to the east.

Ice storms differ from short freezing rain episodes in their duration, total ice accumulation, and areal extent. Ice storms generally last twelve hours or more, accumulate ice several centimetres thick, and affect an area ranging from a few square kilometres to regions as large as several provinces or states. The mean ice storm swath is typically 50 kilometres (30 mi.) wide and 500 kilometres (300 mi.) long. While residents of eastern Canada and the United States may experience some freezing rain each year between late October and mid-April, ice storms that cause severe damage generally warrant major headlines only one year in three. The eastern ice storm "alley" stretches from northern Texas to Newfoundland.

North America's worst ice storms are commonly associated with slow-moving low-pressure systems having very large temperature differences between colliding warm, moist Gulf air and very cold arctic air in their northeastern sector. When these storm systems stall for an extended period over one region, heavy accumulations of ice may blanket a region, causing much destruction. The great ice storm that hit Quebec and New England in 1998 is an extreme example of a stalled storm.

Freezing Rain Formation

Ice storms typically begin with a period of snow and strong easterly winds ahead of the slowly approaching warm front. Advancing warm,

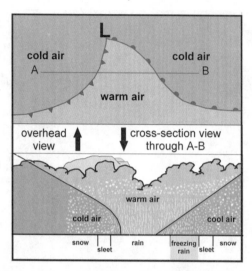

Top view and side view of a winter storm showing area of most likely freezing rain formation.

moist air from the south overruns the surface cold air at the warm front and produces the characteristic vertical temperature layering (warmer air above, colder below) of the pre-warm front atmosphere. As the warm, moist air is forced up, it cools. By the time it reaches the upper cloud levels, its temperature has dropped below freezing, and its liquid moisture forms ice crystals. With additional moisture constantly condensing, the crystals quickly combine into snowflakes.

Now too heavy to be held aloft, the snowflakes fall into the air below. If the lower air temperature is below freezing, the flakes remain frozen and continue their earthward descent as snow. However, if this air layer is above freezing and of proper depth, the snowflakes will melt and continue their fall toward the surface as raindrops. Whether or not freezing rain forms depends critically on the characteristics of the near-surface air layer. If this layer remains above or very near freezing, the raindrops will hit the ground as a cold, but still liquid, rain. If the layer is well below freezing and thick, the rain will refreeze into ice pellets (sleet).

In order to reach the ground as freezing rain, raindrops must be cooled to temperatures below freezing and yet not freeze (a condition known as supercooling) as they fall through a cold surface-air layer before striking the ground or other surfaces. Often small differences in the surface-layer air temperature and raindrop size result in freezing rain mixed with sleet, snow, and non-freezing rain. Glaze ice generally forms when the near-surface air temperature is in the narrow range of –4 to 0°C (25 to 32°F). For thick, destructive glaze formation, freezing rain should fall rapidly as large, slightly supercooled drops,

Aftermath of ice storm, Guelph, Ontario, 1972. (KEITH C. HEIDORN)

and the rate of freezing on the surface must be slower than the rain-fall rate. Once a supercooled water droplet strikes a surface, the impact triggers the transformation of liquid water to ice. When falling drops are small and their freezing rapid, milky hard rime ice forms instead of glaze, its crystalline structure more like sugar than clear ice.

The sensitivity of freezing rain formation to the depth and temperature of the near-surface cold air layer makes precise forecasting of the formation, amount, and ice accumulation rate difficult. Often freezing rain episodes are hit and miss across an area, varying with elevation or degree of urbanization. Ice storms are more uniform and cover larger areas but still pose a challenge to forecasters.

Freezing Rain Impacts

When transportation depended on foot power and beasts of burden, moving along dirt or gravel roads, a glaze storm was generally considered more an inconvenience than a hazard, except perhaps for those who traveled through wooded areas, where falling branches and trees were a danger. Pavement and the automobile brought new travel hazards to both drivers and pedestrians. On relatively smooth horizontal surfaces, such as road pavement and sidewalks, glaze ice lies in rather uniform, smooth slippery sheets that do not easily break when weight is applied. For drivers, the consequences of icing can be serious, as stopping distances on glaze ice are ten times greater than on dry pavement, and double or more that on packed snow.

Power and communication systems using overhead lines are perhaps hardest hit by ice storms. Hanging wire cables may collect ice until the cables break or the rain stops. Diameters of these ice-coated cylindrical cables may reach 5 centimetres (2 in.), adding a weight of 15 to 30 kilograms per metre (10–20 lb./ft.) to the wire. Lines not broken directly may succumb to the combined forces of ice and wind or to falling trees and branches. Even days after the storm has abated, lines may break when they react to the sudden change in their load as the ice falls from them. Vibrations, often violent, may also occur as the ice is shed, snapping weak points in the line under the added strain.

Freezing rain, however, affects more than just human technology. Animals and plants—both wild and domestic—may be killed or injured by ice accumulation. Ice damage rivals disease and insects as destructive agents to trees. Like power cables, tree branches and trunks collect ice in vast quantities. Conifers are the most resistant trees owing to their great flexibility, tapered shape, and lack of trunk branching. A 15-metre (49-ft.) conifer can accumulate as much as 45,000 kilograms (99,000 lb.) of ice during a severe storm.

Deciduous trees exhibit a wide range of vulnerability to ice. Elms and fruit trees are among the most vulnerable, while oak and beech are among the most tolerant of ice loading. In general, younger deciduous trees are better able to survive than older ones due to their more supple nature and limited branching. Multiple large branches in a tree generally lead to more breakage; heavy ice accumulation on a multi-branched trunk may cause a tree to split in two. Icing damages plants by sealing leaves, stems, and buds from the air, thereby suffocating these parts. Similarly, ice sheets formed over snow surfaces, should they persist, may suffocate or poison overwintering species, such as winter wheat.

Animals active in the winter must also cope with an ice storm's effects. Many animals starve when they are unable to reach seeds, buds, or other food locked in the ice. Deer, for example, find it difficult to graze when young shoots become encased in ice. Birds unable to find a sheltered perch during a storm may discover their feet frozen to a branch or their wings covered in ice, rendering them unable to fly. Grouse buried in snowdrifts are often encased by the ice layer and suffocate.

While a walk through a forest after an ice storm may give the impression of large-scale destruction amid crystalline beauty, an ice storm also provides several vital functions. The same mechanical action that downs branches also releases seeds, promotes regeneration, prunes dead or dying branches, and indirectly provides nesting and sleeping cavities for birds and other animals.

Ice storm, Guelph, Ontario, 1972. (KEITH C. HEIDORN)

Afterword

Ice storms and the beauty of glaze ice are not common here in Victoria, though they occur with some frequency each year in the valleys between the mainland coastal mountains, where cold air is trapped and rainfall drops from mild, moist air crossing mountain ridges (a different situation than the frontal storm model characterized in the last chapter). I have had my trying times with ice storms and freezing rain in the past, but the beauty of ice-covered vegetation and terrain still moves me, even if I now see it only in photographs and on television screens.

As the winter solstice approaches, the combination of short days and moisture-laden skies makes most days too gloomy for my tastes, for I am a solar person. Mild, yes, but many times during the Pacific coast winters, I long for the bright skies and crisp, cold airs of the mid-continent winter, even if it would require pulling my lonely parka out of the storage closet. I have come to realize I do not dislike winter extremes, just the requirement to travel in them. Now, with my office in my home, I need not venture forth when conditions deteriorate and can fully enjoy all the beauties of winter conditions.

December mornings continue to arrive later, but the shift to later sunsets is apparent to the keen eye as equation of time factors play with the times of sunrise and sunset. The changes are only minutes a day, but a solar person such as I notices them. As the winter solstice draws nigh and the solar year reaches its ebb, the weather year again comes full circle. For much of North America, save the tropical and subtropical lands, winter weather conditions have arrived and perhaps taken control. The frigid arctic airs continue to migrate out of the dark northern regions, heading for tropical warmth, for winter has just begun in earnest.

By now, I hope your weather eyes and other weather senses have become attuned to the ebb and flow, the cycles of hill and dale, in the daily weather patterns where you live. If so, find a comfortable spot and enjoy the show, for every day tells a new story written across the vault of sky.

Henry David Thoreau wrote in his journals: "For many years I was self-appointed inspector of snowstorms and rainstorms and I did my duty faithfully." I follow in his footsteps, and you can, too. If you have not yet trained your weather eyes, then practice, practice, practice—he says with a smile. You will find the effort enjoyable and rewarding, a lifelong practice to share with friends and family.

In my family, my weather eyes have been passed through my daughter to my three grandsons. When thunderstorms rumble around their mountain BC home, they ask to open the curtains so they can watch the lightning, while their young friends become fearful. They are still too young to understand the science, but I will present that to them in the coming years.

In the dedication to this book, I thanked my kids for putting up with a weird father who enjoyed sitting outside during stormy summer weather and explaining the weather whys to them. I thank also those who have been readers of my Weather Doctor and Science of the Sky Web sites and radio listeners of *The Weather Notebook* who have indicated they enjoyed my contributions there (and from which some of the material herein has been adapted). And the book you hold in your hand would not exist if not for the interest and editorial assistance of Charlene Dobmeier, Meaghan Craven, and Alex Frazer-Harrison of Fifth House Ltd. My thanks to them and those unnamed behind the scenes who made this book possible. And a sincere thank you to my son Derek, who taught me how to use the Internet and code my Web sites. The best parts of the Weather Doctor's site design are his.

Having finished this project, I plan to reward myself with some extra time on my balcony or in front of my window, comfortable in my easy chair, with many hours of field work tossed in. Perhaps tomorrow, Island View Beach or the summit of Mount Douglas will beckon.

My weather eyes are still looking up, as they do every day.

Glossary

This glossary defines terms not fully defined in the book.

air mass: a large dome of air that has similar horizontal temperature and moisture characteristics and often a rather homogeneous change of temperature with height

alto: a prefix to cloud-type names for clouds generally found between 3,000 and 7,000 metres (9,800 and 22,000 ft.)

Bermuda High: a semi-permanent region of high pressure found over the western subtropical North Atlantic Ocean near the island of Bermuda. It produces southerly airflows that often move warm, humid air into the eastern regions of North America.

condensation nuclei: liquid or solid particles, such as acidic droplets or dust, that provide a surface upon which water vapor can condense into cloud droplets or form ice crystals

cumulonimbus: tallest of all clouds, many extend to over 18,000 metres (60,000 ft.); precipitation, often in the form of thundershowers, is associated with them

cumulus humilis: small cumulus clouds with flat bottoms and slightly rounded tops of small vertical extent

cyclogenesis: the process of formation of a cyclone, or the intensification of cyclonic (counterclockwise) flow around a low-pressure system

deposition: the appearance of ice on a surface forming directly from water vapor

glaze ice: transparent and homogeneous ice forming on vertical and horizontal surfaces by the freezing of supercooled water

inversion: a layer of air in which the temperature increases with height. Meteorological convention considers temperature decreasing with height as the norm, thus when temperature decreases with height, it is inverted.

lake effect: the effect of any lake in modifying the weather and climate along its shore and a distance inland, which depends on the lake's size

latent heat: heat released or required when matter changes phase. The latent heat of fusion applies to a change from a liquid to a solid; the latent heat of condensation, to a change from gas to liquid.

mesoscale: space scale, also known as the local scale, that covers atmospheric elements from a few kilometres to tens of kilometres in the horizontal dimension and from the surface to the top of the planetary boundary layer, about 1 kilometre (3,300 ft.) in depth

microscale: space scale that includes all atmospheric processes less than a few kilometres in size

occluded front: complex frontal region where warmer air lies over the zone separating cold air from cool air, usually occurs when a cold front overtakes a warm front and lifts it

pressure gradient: the horizontal change of pressure that occurs over a fixed distance at a fixed altitude

ridge: an elongated area of high pressure around which the winds circulate in a clockwise direction in the northern hemisphere

sea-level pressure: the atmospheric pressure measured at sea level or measured at elevation and adjusted to its sea-level equivalent. Used to draw pressure fields on surface weather maps.

solar declination: the sun's noon angle relative to the equator. It ranges from 23.5° north on the summer solstice to 23.5° south on the winter solstice and is 0° on the equinoxes.

solar season: one of four three-month periods that divide the year for non-tropical latitudes, based on day length or potential daily sunlight

squall line: a non-frontal line of vigorous thunderstorms with strong, gusty winds and usually heavy precipitation

sublime or sublimation: the transition of a substance from the solid phase directly to the vapor phase without being in the liquid phase. The opposite of sublimation is deposition.

supercooled: a state when the temperature of liquid water falls below 0°C (32°F) without freezing

synoptic scale: space scale that covers weather elements such as high and low pressure systems, air masses, and frontal boundaries—features commonly found on standard weather maps

thermal: a rising parcel of warm, and less dense, air generally produced when Earth's surface is heated or when cold air moves over a warmer surface such as warm water

trough: an elongated area of low atmospheric pressure

unstable: atmospheric state that arises when a rising air parcel becomes less dense than the surrounding air. Since the parcel's temperature will not cool as rapidly as the surrounding environment, it will continue to rise and produce strong updrafts.

wind shear: the rate of wind speed or direction change with distance. Vertical wind shear is the rate of change with respect to altitude.

Bibliography

Ahrens, C. Donald. *Meteorology Today*. St. Paul: West Publishing Company, 1991.

Bentley, W. A., and W. J. Humphreys. *Snow Crystals*. New York: Dover Books, 1931.

Blanchard, Duncan C. *From Raindrops to Volcanoes: Adventures with Sea Surface Meteorology* (reprint). New York: Dover Publications, 2004.

Blanchard, Duncan C. *The Snowflake Man: A Biography of Wilson A. Bentley*. Blacksburg, Virginia: The McDonald & Woodward Publishing Company, 1998.

Day, John A. *The Book of Clouds*. New York: Silver Lining Books, 2002.

Greenler, Robert. *Rainbows, Halos and Glories*. Cambridge: Cambridge University Press, 1980.

Hamblyn, Richard. *The Invention of Clouds: How an Amateur Meteorologist Forged the Language of the Skies*. New York: Farrar Straus & Giroux, 2001.

Henson, Robert. *The Rough Guide to Weather*. London/New York: Rough Guides/Penguin, 2002.

Lee, Raymond L., Jr., and Alistair B. Fraser. *The Rainbow Bridge: Rainbows in Art, Myth, and Science*. University Park, PA: Pennsylvania State University Press, 2001.

Libbrecht, Kenneth, and Patricia Rasmussen. *The Snowflake: Winter's Secret Beauty*. Stillwater, MN: Voyageur Press, 2003.

Lynch, David K., and William Livingston. *Color and Light in Nature*, 2nd ed. Cambridge: Cambridge University Press, 2001.

Mergen, Bernard. *Snow in America*. Washington: Smithsonian Institution Press, 1997.

Minnaert, M. *The Nature of Light and Color in the Open Air*. New York: Dover Publications, 1954.

Nese, Jon M., and Lee M. Grenci. *A World of Weather: Fundamentals of Meteorology*, 3rd ed. Dubuque, IA: Kendall/Hunt Publishing Company, 2002.

Phillips, David. *Blame It on the Weather: Strange Canadian Weather Facts*. Toronto, Ontario: Key Porter Books, 1998.

Phillips, David. *The Day Niagara Falls Ran Dry*. Toronto, Ontario: Key Porter Books, 1993.

Schaefer, Vincent J., and John A. Day. *A Field Guide to the Atmosphere*. Boston: Houghton Mifflin Company, 1983.

Watson, Lyall. *Heaven's Breath*. New York: William Morrow and Company, Inc., 1984.

Williams, Jack. *The Weather Book*. New York: Vintage Books, 1997.

Index

Index

FIFTH
HOUSE

About Fifth House

Fifth House Publishers, a Fitzhenry & Whiteside company, is a proudly western-Canadian press. Our publishing specialty is non-fiction as we believe that every community must possess a positive understanding of its worth and place if it is to remain vital and progressive. Fifth House is committed to "bringing the West to the rest" by publishing approximately twenty books a year about the land and people who make this region unique. Our books are selected for their quality, saleability, and contribution to the understanding of western-Canadian (and Canadian) history, culture, and environment.

Look for the following Fifth House titles at your local bookstore:

The BC Weather Book: From the Sunshine Coast to Storm Mountain by Keith C. Heidorn

The Canadian Weather Trivia Calendar by David Phillips (published annually)

But It's a Dry Cold!: Weathering the Canadian Prairies by Elaine Wheaton